Glucose Polymers in Health and Disease: The Role of Caloreen

Glucose Polymers in Health and Disease: The Role of Caloreen

Edited by
N. P. Mallick
B.Sc., M.B., Ch.B., F.R.C.P.

Proceedings of a Symposium held at the University of Manchester, England, June 1976

MTP

Published in the UK by:

MTP Press Limited,
St. Leonard's House,
Lancaster, Lancs
England

First published 1977

ISBN 0 85200 182 7

Printed in Great Britain at The Spottiswoode Ballantyne Press
by William Clowes and Sons, Limited, London, Colchester and Beccles

Contents

List of Contributors

S. P. ALLISON: *Nottingham General Hospital*

S. T. ATHERTON: *Whiston Hospital, Prescot, Merseyside*

H. K. BERRY: *Department of Pediatrics, University of Cincinnati College of Medicine, USA*

E. M. BOOTH: *Dietetic Department, Manchester Royal Infirmary*

D. DAVIES: *Department of Endocrinology, Manchester Royal Infirmary*

D. E. M. FRANCIS: *Dietetic Department, Hospital for Sick Children, London*

M. M. HUNT: *Department of Pediatrics, University of Cincinnati College of Medicine, USA*

R. L. INGBERG: *Department of Pediatrics, University of Cincinnati College of Medicine, USA*

F. W. KELLOGG: *Department of Pediatrics, University of Cincinnati College of Medicine, USA*

N. P. MALLICK: *Department of Renal Medicine, Manchester Royal Infirmary*

D. B. McWILLIAM: *Whiston Hospital, Prescot, Merseyside*

C. R. RICKETTS: *MRC Industrial Injuries and Burns Unit, Birmingham*

M. ROBINSON: *Whiston Hospital, Prescot, Merseyside*

E. SHERWOOD JONES: *Whiston Hospital, Prescot, Merseyside*

A. M. J. WOOLFSON: *Nottingham General Hospital*

D. M. WRIGHT: *Whiston Hospital, Prescot, Merseyside*

1

Introduction

J. MILNER and N. P. MALLICK

The story of the glucose polymer Caloreen began in Manchester and we thought it appropriate that we should meet here to put together the data we had collected from experimental and clinical experience in the UK and USA. Sick people require energy giving foods and calories derived from sugars are important to them. There are many problems in determining the way in which foods are utilised in the seriously ill and no doubt there will, in time, be new insights which will help our understanding.

It seems clear that Caloreen has proved of value in a wide range of diseases although it was in the renal field that it first found a practical place. The very ill patient has difficulty in taking food and any calorie source should be either very tasty or quite without taste. It should be freely miscible with water and so easily added to many foods; it should not present the intestine with a large osmotic load which might cause vomiting or diarrhoea; it should be free of electrolytes and protein, and should be as readily utilised in sickness as in health.

In retrospect the use of glucose polymers for this purpose seems obvious and it could be predicted that when given by mouth they would be metabolised to provide energy in the form of monosaccharide. This proved to be the case and the papers in this Symposium show that Caloreen has been of benefit in the treatment of acutely ill patients and in a variety of diseases, which has exceeded our initial expectations.

It was obviously necessary to investigate the polymer composition in more detail and to establish how it was metabolised. It was also clear that a utilisable polysaccharide might prove important as a source of intravenous calories. All these aspects of glucose polymers were studied co-operatively by a group of workers in Manchester, Liverpool, Birmingham and Nottingham. Reports from each of these centres form the basis of papers in this Symposium. As a result of our studies we have established guidelines for the ideal form of glucose polymer for intravenous clinical use.

This Symposium deals particularly with the use of Caloreen in disease. Its greatest contribution may yet lie in combating the problem of calorie deficiency in disease processes throughout the world.

2

Sugars and Dextrins for Dietary Use

C. R. RICKETTS

We eat to provide energy for movement and to help the incorporation of amino acids into tissue protein. Normally an adult needs about 3000 calories per day. An injured patient with a high metabolic rate needs much more. About half this energy can come from fat and half from carbohydrate; 300 or 400 g of carbohydrate is normally taken, most of which is starch and sucrose with a little lactose.

Starch, sucrose and lactose are not normally absorbed unaltered by the small intestine. Digestion by hydrolysis to the constituent monosaccharides is an essential step in their utilisation. Although the amylase of saliva plays some part, the major part of the digestion is carried out in the lumen of the small intestine by pancreatic amylase which only hydrolyses the 1:4 link in starch, producing maltose, together with some maltotriose and limit dextrins. Another intestinal enzyme splits the 1:6 links enabling complete conversion to maltose. Also present in the small intestine at this stage are the unchanged sucrose and lactose derived from food.

The brush border contains several disaccharidases which continue digestion. Maltase acts on maltose and maltotriose, producing glucose; sucrose is converted to glucose and fructose, and lactose is converted to glucose and galactose. The monosaccharide mixture presented to the cells of the intestinal wall for absorption has the approximate composition 80% glucose, 15% fructose and 5% galactose.

It seems likely that the disaccharidases of the brush-border membrane of the mucosal cells of the small intestine are orientated in such a way that the disaccharides are split and the resulting monosaccharides absorbed in close sequence. This integration of digestion and absorption is a fairly recent concept and seems to be a general feature, applying also to proteins and fats.

If for any reason disaccharide hydrolysis does not occur, clinical symptoms, for example, diarrhoea, may occur. This has been attributed to two main causes, namely the osmotic effect of unabsorbed disaccharide and its conversion by colonic bacteria to organic acids, further increasing the osmotic and irritant effects.

Refined sugars, like sucrose, glucose, fructose and sorbitol, when eaten in quantity are, in a sense, 'unnatural foodstuffs'. Starch, or some predigested form of starch such as dextrin, provide the source of energy from carbohydrate to which we are best adapted.

The composition of some glucose and dextrin preparations can be investigated by gel filtration on a column of Sephadex G-25. Small dextrin molecules can diffuse right into the spherical grains of Sephadex and so their passage through the column is delayed; large molecules cannot enter the grains and must go round them; these large molecules come off the column first. The separation is recorded with automatic measuring equipment, utilising the refractive index of the solution leaving the column.

Liquid glucose, a widely used material, has the nominal composition of 19% glucose, 14% maltose with a variety of higher saccharides, mainly less than 8 glucose units in length (Table 1). Figure 1 shows

Table 1 Composition of liquid glucose

Glucose
Maltose (including isomaltose)
Trisaccharides
Tetrasaccharides
Pentasaccharides
Hexasaccharides
Heptasaccharides
Octa and higher saccharides

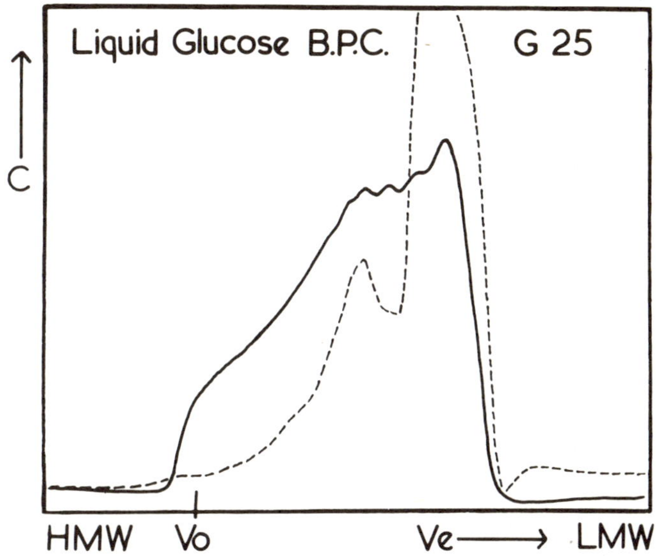

Figure 1 Liquid glucose: Sephadex G-25 separation of liquid glucose

these components graded by Sephadex G-25 separation according to their molecular size (solid line), the largest—maltodextrins—on the left and the smallest—glucose—on the right. The dotted line shows the effect of salivary amylase in breaking down the dextrins to form maltose. There is incomplete conversion to maltose, as shown by the small peak remaining in the approximate position one would expect of a pentasaccharide. This amylase-resistant material probably contains the branching (1:6 link) component of the original starch molecule. The intestine, and also many other tissues of the body, contains the enzyme maltase which breaks maltose into two molecules of glucose, but only the intestine is known to contain an enzyme able to split the 1:6 link.

Figure 2 shows the dextrins present in the proprietary product Caloreen separated in the same way according to their molecular size. Note that there are more of the large molecules, as shown by the peak in the void volume of the column. These molecules are too large to enter into the Sephadex grains. After incubation of Caloreen with amyl-

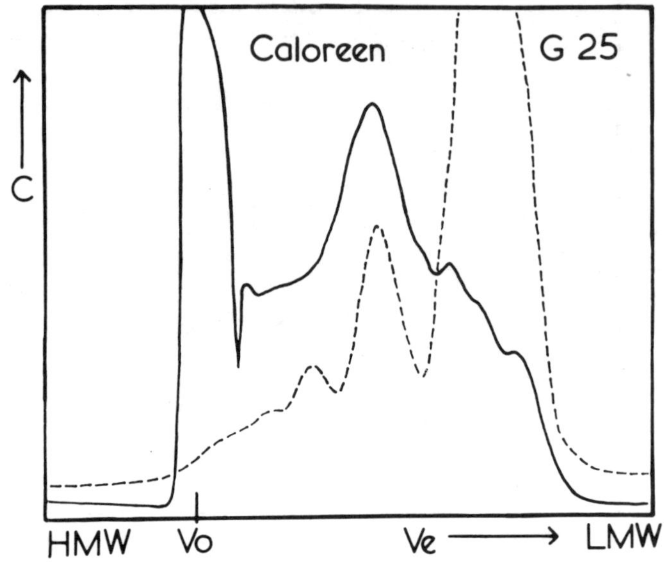

Figure 2 Sephadex G-25 separation of Caloreen

ase the dotted curve shows that the larger dextrin molecules have been converted to maltose but again a small proportion is resistant to amylase; presumably other enzymes in the intestine break the branch linkage enabling complete conversion to maltose to proceed.

An important feature of these dietary carbohydrates is their average molecular weight, because the higher the molecular weight the lower the osmotic pressure of a solution containing a given number of calories. For example a 20% solution of dextrose, liquid glucose, or Caloreen each contains 800 calories per litre but each has a different osmotic effect (Table 2). The ingestion of a large volume of strongly hypertonic solution may lead to drawing of excessive amounts of water into the intestine from other tissues, with consequent metabolic disturbances.

We have considered the theoretical possibility of giving these dextrins intravenously to patients who are unable to take them by mouth. Most of the molecules in the foregoing preparations are smaller than the renal threshold for molecular size. This is illustrated in Figure 3 which shows the well-known infusion material Dextran 40

Table 2 Osmotic effect of various carbohydrates (20% solution)

	Milliosmoles per litre	*Isomotic concentration (%)*
Dextrose	1110 (approx.)	5
Liquid glucose	571	10.1
Caloreen	240	24.1

separated by gel filtration on a column of Sephadex G-200 according to the size of the molecules, the largest on the left and smallest on the right. Some 68% of Dextran 40 is of sufficiently small molecular size to pass through the glomerular basement membrane into the urine. It is also clear that all the molecules of Caloreen are smaller than the threshold for renal filtration.

After intravenous injection into volunteers, early urine samples contained dextrin with a similar distribution of molecular weight to

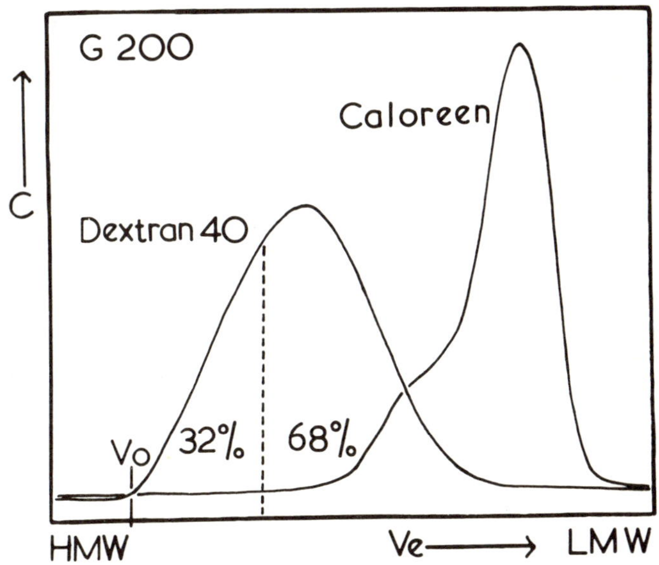

Figure 3 Sephadex G-25 separation of Dextran 40

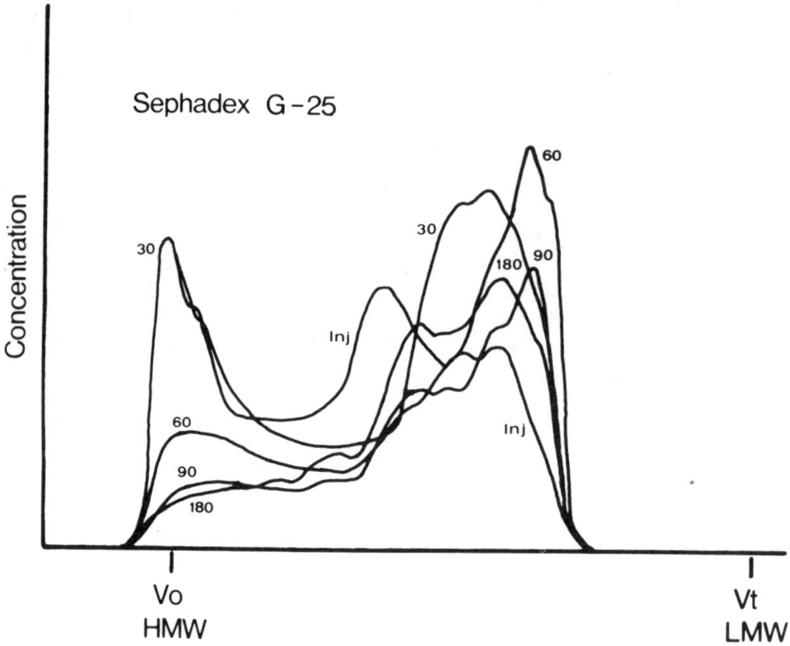

Figure 4 Sephadex G-25 separation of early urine samples after intravenous bolus injection of Caloreen (inj = injected material). Time of each sample tested indicated in minutes

the injected dextrin (Figure 4). Later samples, 2 or 3 hours after injection were found to contain a higher proportion of larger molecules. This finding is an indication of the breakdown of dextrin by action of the enzyme amylase in the bloodstream.

Continuous infusion of dextrin at 15 g/h for 3 hours led to a slow build up of carbohydrate in the bloodstream. Unexpectedly, the plasma amylase level fell during the infusion. Experiment showed that the presence of dextrin did not interfere with the estimation of plasma amylase. Possibly the 1:6 branch links in the dextrin which are known to impede amylase action, led to a complex of amylase with dextrin being taken up by the reticulo-endothelial cells, and so removed from the bloodstream.

It was logical to examine the effect of injecting maltose (see also p. 80), the main product of the action of serum amylase on dextrin.

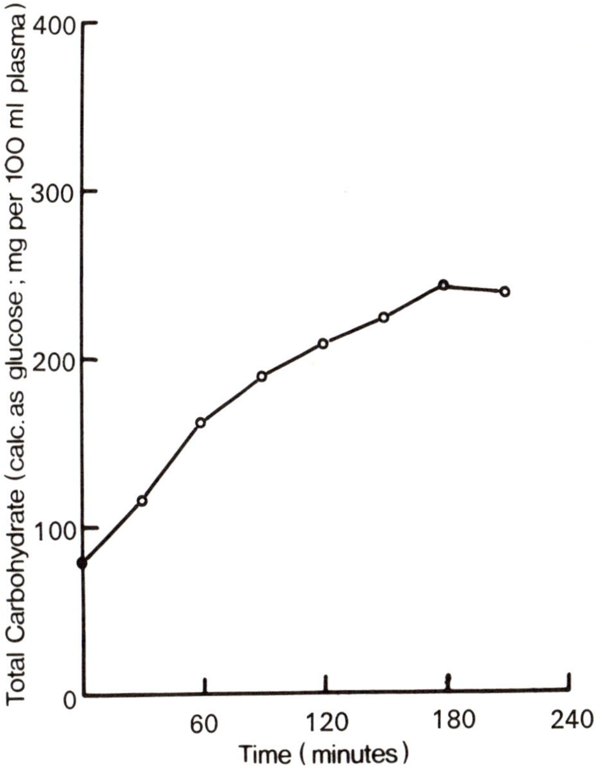

Figure 5 Continuous infusion of maltose 15 g/h

Continuous infusion of maltose at 15 g/h for 3 hours resulted in a rising level of blood carbohydrate (Figure 5) but only a very small amount of carbohydrate was excreted. Infusion of maltose was accompanied by a dramatic increase in the respiratory quotient of a fasting subject. Such an increase was unobtainable by a continuous infusion of dextrin presumably because of slow conversion to maltose.

Discussion

Dr Mallick—*We have shown that Caloreen given orally is absorbed and metabolised as quickly as oral glucose itself. However the mixed composition of Caloreen is a potential problem if the intravenous route is used, as Dr Ricketts showed.*

Mr Milner—*Could we talk a little about enzymes in the gut? We hear a lot about the saliva and pancreatic amylase but little about intestinal amylase.*

Dr Ricketts—*The human body is a little like a Rolls Royce in that there are two of most things that are vital. There is a source of amylase in the saliva which mixes with food and passes into the intestine. All the amylases in the body are alpha-amylases and carry out random splitting of the starch molecule down to small fragments which might be maltose with two units, trisaccharides with three units and so on. This is in contrast to beta-amylase which is a plant enzyme and splits off glucose units in pairs from the end of the starch molecule. As far as I know beta-amylase is not present in mammals.*

Dr Davies—*You mentioned serum amylase falling in patients infused with Caloreen and explain this by the adsorption of the amylase into the molecule and its uptake by the reticulo-endothelial system. Could an alternative explanation be that amylase has been utilised to break down the Caloreen in circulation?*

Dr Ricketts—*Yes, but an enzyme is not really 'utilised' in that sense; an enzyme is a catalyst and still present when the reaction is complete.*

Dr Berry—*Would you comment on the role of amylopectin which, after all, is a major part of most starches?*

Dr Ricketts—*Yes, the intestine has an enzyme capable of splitting the 1:6 branching links in amylopectin so that it is metabolised in the intestine. When such 1:6 linked compounds are given intravenously, this does not occur.*

Mr Milner—*The gut contains bacteria, so what is the role of the intermediates of carbohydrates in relation to these bacteria?*

Dr Ricketts—*We have been looking at the growth of bacteria on Caloreen and, in general, bacteria do not grow on it. When one adds peptone to the medium, providing a source of nitrogen, then some intestinal bacteria, which are able to split starch-like molecules, grow quite well.*

Dr Sherwood-Jones—*I am interested in this 1:6 splitting enzyme which you passed over rather quickly. Is this an extracellular enzyme, is it in the brush-border cell wall or even intracellular?*

Dr Ricketts—*I think it is an intracellular enzyme; in some plants it is called a debranching enzyme; this same enzyme for debranching starch molecules occurs in animal sources but very little has been extracted and I am not aware of extensive work on it. However, it must be present in the intestinal wall, since glycogen has 1:6 links and can be completely metabolised.*

Mr Milner—*I think the problem is that there is insufficient classification of animal amylases. The serum of some animals can* *convert Caloreen rapidly to maltose. A dog can convert intravenous Caloreen at ten times the rate that man can achieve. We have looked at about eight animal species and in all of these there was rapid conversion of intravenous Caloreen to maltose and glucose.*

Dr Ricketts—*We should remember that the presence of amylase in circulation is a mere accident, as it got there by breakdown of a cell somewhere else in the body.*

Summaries

Sugars and Dextrins for Dietary Use

C. R. Ricketts

Caloreen consists of a number of glucose units of differing chain length linked either at the 1:4 or 1:6 position. The brush border of the small intestine contains enzymes acting on each of these links, so that Caloreen is broken down to monosaccharides and probably all absorbed as such. By Sephadex gel separation the saccharide composition of Caloreen has been compared to that of 'liquid glucose' and Dextran. After intravenous injection of Caloreen, Sephadex separation of urinary samples shows that, with time, some of the 1:4 linked saccharides appear to be metabolised.

Sucres et Dextrines à Usage Diététique

C. R. Ricketts

Le Caloreen est constitué de plusieurs motifs à base de glucose ayant des longueurs de chaîne différentes unis en position 1:4 ou 1:6. La bordure en brosse de l'intestin grêle renferme des enzymes agissant sur ces liaisons, si bien que le Caloreen est fragmenté en monosaccharides et vraisemblablement totalement absorbé sous cette forme. Par séparation sur gel de Sephadex, on a comparé la composition en saccharides du Caloreen à celle du produit d'hydrolyse partielle de l'amidon et du Dextran. Après injection intraveneuse de Caloreen, la séparation sur Sephadex d'échantillons d'urines montre qu'au cours du temps une partie des saccharides à liaisons 1:4 semble être métabolisée.

Zucker und Dextrine als Diätmittel

C. R. RICKETTS

'Caloreen' besteht aus einer Anzahl Glukose-Einheiten unterschiedlicher Kettenlänge, die in der 1:4- oder 1:6-Stellung verknüpft sind. Die Darmzotten des Dünndarms enthalten Enzyme, die auf jede dieser Bindungen so einwirken, daß 'Caloreen' in Monosaccharide gespalten wird, die wahrscheinlich alle als solche resorbiert werden. Mittels Sephadexgel-Trennung ist die Saccharidzusammensetzung von 'Caloreen' mit der von 'flüssiger Glukose' und Dextran verglichen worden. Nach intravenöser 'Caloreen'-Injektion ergibt sich aus der Sephadex-Trennung der Harnproben, daß in Abhängigkeit von der Zeit einige der 1:4-verknüpften Saccharide abgebaut zu werden scheinen.

Azúcares y dextrinas para Uso Dietético

C. R. RICKETTS

Caloreen está constituido por unidades de glucosa de diferente longitud de cadena unidas entre sí en la posición 1:4 ó 1:6. El borde en cepillo del intestino delgado contiene enzimas que actúan sobre cada una de estas uniones, de manera tal que Caloreen es convertido a monosacáridos, y posiblemente es absorbido totalmente en esa forma. Por separación a través de columna de 'Sephadex' la composición sacárida del Caloreen ha sido comparada a la de la 'glucosa líquida' y el Dextran. Luego de la inyección intravenosa de Caloreen la separación por 'Sephadex' de las muestras urinarias muestra que, con el transcurso del tiempo, algunos de los sacáridos con uniones 1:4 parecen haber sido metabolizados.

3

The Dietetic Role of Caloreen

E. M. BOOTH

This paper is intended to outline the use of Caloreen in general dietetics. The dietitian/patient relationship has become an integral part of medical care in a number of fields. Having been involved from The outset in the search for a high-energy supplement which was free of taste, flavour, nitrogen and electrolytes and more recently, in the discussions on the possibility of using such a source intravenously, this story has come to have a certain fascination for me. Unless the patient could take and continue to take the supplement which evolved, and this, of course, was Caloreen, the exercise was worthless.

Caloreen is a fine white powder slightly hygroscopic and sticky. It is flavourless and has minimal sweetness; it dissolves to produce a highly concentrated solution which can be added to other items of food and drink. Its osmolality is considerably less than other carbohydrate solutions. It yields approximately 4 kilocalories/g and is available in two forms in Britain—Caloreen, which is virtually electrolyte free and Gastro-Caloreen which has not been processed to remove the electrolytes. It can be prescribed as a Borderline Substance for chronic renal failure and other conditions requiring high calorie, low fluid, low electrolyte diets.

The success of Caloreen is due to its small bulk and minimal sweetness. As a carbohydrate source in catabolic states it provides a valuable boost to energy intake, thereby sparing protein. Because it is less hypertonic than other sugar solutions, Caloreen does not induce or aggravate vomiting and diarrhoea. The original product was manufactured for patients with *chronic renal failure*. The Giordano–Giovannetti diet containing minimal essential amino acids had been formulated but required a high energy intake to be effective. Suitable patients were wasted but anorexic. Most were vomiting and others had sore mouths. If patients could be persuaded to take at least 2500 kilocalories their biochemical state improved, as did their morale. Low protein bread had never been totally acceptable to these patients so Caloreen provided an answer.

Mixed into different dishes or combined with fats such as cream, butter or Prosparol, an oil and water emulsion, recipes for such foods as icecream or cream-filled brandy snaps and other confectionery were concocted. Dietetic support and encouragement were vital to ensure that patients and their relatives could carry out the therapy. Because Caloreen is a dry, electrolyte-free powder it is eminently suitable where renal or hepatic diseases are accompanied by water and electrolyte retention. In *acute renal failure* the properties of Caloreen make it extremely useful for even in concentrated 'dry' diets, Caloreen is unlikely to aggravate a sore mouth or cause vomiting or diarrhoea.

The management of the *nephrotic syndrome* provides another nutritional challenge; the prime problem here is hypoproteinaemia, complicated by anorexia. Dietary intake in excess of 100 g of protein is often required. Knowing the importance of protein a patient may struggle through breakfast containing two eggs, a sandwich lunch with meat or cheese equivalent, and then tackle a steak or two chops for his evening meal, but this is probably at the expense of other foods, because he is anorexic. As a result the protein will be metabolised for energy, and meanwhile the patient becomes more protein depleted. Initially treatment would be with small meals and a high caloric supplement, which should be presented in a palatable form. Examples would be an iced milk shake containing Caloreen, an oil emulsion, flavouring and a protein supplement or amino acids. A 300 ml milk shake of this type can be laced with all the ingredients mentioned to produce 46 g of protein and nearly 700 kilocalories (Table 1). This type of mixture can reverse the wasting by providing both nitrogen and energy. With improvement and return of appetite the patient can

Table 1 Milk shake

		Protein	*Kilocalories*
1 packet	Carnation Build-up	8	124
300 ml	milk	10.5	198
30 g	Casilan	28	108
30 ml	Prosparol		130
30 g	Caloreen		120
		46.5 g	680

resume a varied diet covering the protein deficit, and Caloreen supplements. In this way the improvement is maintained. *Acute hepatic failure* is an instance where energy in the form of carbohydrate is needed to protect the damaged organ. It is extremely simple to mix 500 g of Caloreen in 1 or 2 litres of diluted fruit juice, thereby providing 2000 kilocalories. Group B vitamins are needed to help metabolise a high carbohydrate intake of this kind. Protein requirements can be met, bearing in mind that one egg in 200 ml of milk provides 14 g of protein and satisfies the patient's minimal essential amino acid requirement for 1 day. Indeed, in all hepatic diseases, Caloreen has a useful role to play as a carbohydrate concentrate. If it is not necessary to restrict electrolytes, the less processed and therefore cheaper, Gastro-Caloreen, is preferred.

Gastro-Caloreen has found a special place in paediatrics, for example as a carbohydrate source in *cystic fibrosis*, in specific forms of carbohydrate intolerance and in the synthetic diets occasionally necessary, alas, until Medicine provides a more sophisticated method of treatment. With its lack of detectable flavour, Gastro-Caloreen has led many dietitians to the art of deception in the treatment of *anorexia nervosa*, especially in its early stages. Cunning as these patients invariably are and armed with considerable knowledge of nutrition, their management presents a mammoth problem to both physician and dietitian. Yoghurt is almost invariably acceptable to this group of emaciated individuals; strongly fortified with large amounts of Gastro-Caloreen, cream or Prosparol it will usually be eaten twice a day. In this way, each carton of yoghurt provides 370 kilocalories thus making a good additional calorie source.

Nasogastric tube feeding is renowned for producing both diarrhoea and vomiting, adding considerably to undernutrition nursing care and laundering. Two very simple procedures seem to eliminate the problems almost completely. Firstly, simple dilution of the feeds with water to approximately 2 or $2\frac{1}{2}$ litres and secondly, substitution of Gastro-Caloreen for the glucose content of feeds.

In conclusion, the simplicity of Caloreen, both in its use and presentation, has filled a gap in the feeding of the emaciated patient, producing in its wake a return of appetite. Even as a supplement in tea, coffee and milk drinks it has a role to play in the management of many conditions: pre- and post-operatively, in youth and old age, in medicine and surgery. So often when a high protein intake is prescribed for the patient, it is a high energy intake that is required to ensure that the nitrogen can be used effectively.

Discussion

Dr Mallick—*From your unrivalled experience, what are the best ways to provide Caloreen in a normal diet?*

Miss Booth—*I think probably the simplest way is in a cup of tea or coffee. Two heaped teaspoons well stirred will give over 100 kilo-calories. We also make a concentrated solution and add it to fruit juices, soups or savouries because the sweetness is not detectable. Then it can be added, with little sweetness, in baking.*

Mr Milner—*What about raising the pyridoxine level in patients who are getting a high calorie intake. Could we improve in any way the conversion of intermediate carbohydrates to amino acids?*

Dr Allison—*We have very little knowledge of the vitamin require-ments of people who are ill, and we tend to blast them with excess, adding B group vitamins to the food supplements we give.*

Mr Milner—*In the nephrotic syndrome where protein production in the liver is greatly increased, a study of the role of pyridoxine would be interesting.*

Dr Mallick—*It is true that the metabolic state of the patient with nephrotic syndrome has not been studied in a disciplined way for 23 years.*

Dr Allison—*The fundamental principle is that the amount of vitamin*

B we need to give is not very different in one nutritional situation or another. I think that the subject of nutrition has been beset by compartmentalisation, and that we have looked at particular diseases and regarded the 'physiology' of each disease as something peculiar and divorced from the rest. This has not been a useful approach.

Dr Davies—*We have been studying patients with 'brittle diabetes'. Recently we were unable to obtain adequate control in one of my patients by increasing her insulin because this merely increased the swings of blood sugar. She was losing weight, and had features of starvation ketosis with ketonuria and ketonaemia and a high plasma triglyceride level, which made her serum quite milky. We decided to give her Caloreen as a dietary supplement. She tolerated this very well and within 48 hours she had a clear plasma, and her diabetes was surprisingly simple to control. Of course, we increased the insulin to cope with the dietary intake which we increased to 250 g of carbohydrate a day and this resulted in sufficient improvement to stop her weight loss. She was now able to return home.*

Dr Allison—*We had a similar patient; I think that this is an illustration of the old adage that fat is burned in the fire of carbohydrate. For fat to participate in Kreb's cycle, carbohydrate is necessary.*

Miss Bateman—*We too use a great deal of Caloreen for our renal patients and in our tube feeds. We also use it for patients with severe liver disease in whom we have to restrict protein intake. We hope by giving them Caloreen to prevent them breaking down their own body tissue, and so minimise the chance of hepatic encephalopathy. We give Caloreen in fruit juice to try and give potassium at the same time. The main difficulty I find with Caloreen is not that it is sweet but that if you make too concentrated a solution you increase the viscosity of the liquid so much that it becomes quite unacceptable to the patient.*

Summaries

The Dietetic Role of Caloreen

E. Booth

Caloreen is a valuable adjunct in dietetics because it provides a high energy supplement free of taste and flavour, nitrogen and electrolytes. It can be added to many foods, and used in a variety of conditions such as chronic renal failure, nephrotic syndrome, acute hepatic failure and anorexia nervosa. It is particularly useful in nasogastric tube feeds for the seriously ill since its low osmolarity provides calories without a significant osmotic load. Gastro-Caloreen, from which electrolytes have not been removed, is an alternative preparation which can be used when there is no need to control electrolyte intake.

Rôle Diététique du Caloreen

E. Booth

Le Caloreen est un aliment utile en diététique car il constitue un supplément énergétique important sans saveur ni odeur ne contenant ni azote ni électrolytes. On peut l'ajouter à de nombreux aliments et on l'utilise dans de nombreux cas tels que l'insuffisance rénale chronique, le syndrome néphrotique, l'insuffisance hépatique aiguë et l'anorexie mentale. Il est particulièrement utile pour l'alimentation par sonde gastrique nasale dans le cas de maladies graves car sa faible osmolarité apporte des calories pratiquement sans surcharge osmotique. Le Gastrocaloreen, dont on n'a pas éliminé les électrolytes, constitue une autre préparation que l'on peut utiliser lorsqu'il n'est pas nécessaire de limiter l'apport d'électrolytes.

Die diätetische Bedeutung von 'Caloreen'

E. BOOTH

'Caloreen' ist bei Diätetika eine wertvolle Ergänzung, weil es eine energiereiche Versorgung bietet, die von Geschmack und Aroma, Stickstoff und Elektrolyten frei ist. Es kann vielen Nahrungsmitteln zugesetzt und bei einer Vielzahl von Befunden, wie chronisches Nierenversagen, nephrotisches Syndrom, akutes Leberversagen und Anorexia nervosa, verwendet werden. Es ist bei der Nasen/Magen-sondenernährung von Schwerkranken besonders nützlich, da seine niedrige Osmolarität Kalorien ohne wesentliche osmotische Belastung liefert. 'Gastrocaloreen', bei dem die Elektrolyte nicht entfernt sind, ist ein alternatives Präparat, das eingesetzt werden kann, wenn keine Kontrolle der Elektrolytzufuhr erforderlich ist.

Rol del 'Caloreen' en Dietética

E. BOOTH

Caloreen constituye un valioso agregado al campo dietético, dado que suministra un suplemento de alto valor calórico carente de sabor y aroma, desprovisto de nitrógeno y electrolitos. Puede ser agregado a numerosos alimentos, y utilizado en una variedad de estados tales como insuficiencia renal crónica, síndrome nefrótico, insuficiencia hepática aguda y anorexia nerviosa. Resulta particularmente útil cuando la alimentación debe efectuarse por sonda nasogástrica en pacientes gravemente enfermos, ya que por su baja osmolaridad aporta calorías sin producir una carga osmótica significativa. El Gastro-Caloreen es un preparado que conservr los electrolitos, y por ende puede ser utilizado cuando no es necesario el control de la ingestión electrolítica.

4

Caloreen and Its Use in Sick Children

D. E. M. FRANCIS

This paper concerns the use of energy supplements for children in various clinical conditions. Children require a balanced diet containing the essential nutrients to keep well and growing; they have different nutritional requirements when unwell. All infections cause a certain amount of catabolism. In most children with the acute infections of childhood this is unimportant but becomes relevant especially in the child where one of the energy-producing nutrients in the diet has to be restricted because of the underlying disease. Our aim is good nutrition, good health and good growth with reserves of nutrients so that the child can keep well nourished.

Protein, fat and carbohydrate all contribute to energy needs. Thus energy supplements are needed if we have to restrict fat as in some of the lipid disorders, protein in chronic renal failure or even sodium, for in restricting sodium a supplement may be required to replace bread which is forbidden because of its salt content. In this case the primary aim is not to restrict a food substance; this results coincidentally, reducing the energy value of the diet.

Caloreen is a useful tool in dietetics, but it is not the only one; for instance children will usually accept sweets very well. Of course a dietitian advises regarding dental hygiene but an acceptable energy supplement is the first priority. Other useful calorie sources are potato crisps and these may be cooked in MCT oil for the child who has to have the half-digested medium-chain triglyceride fats; protein-free bread; aproten spaghetti; Caloreen or Gastro-Caloreen.

Natural foods should be encouraged wherever possible. For example giving chocolate as part of a measured diet is much better than relying solely on pure carbohydrate since the chocolate is easily accepted and also contributes some trace minerals and vitamins. Combining products such as protein-free flour, fat and Caloreen to make palatable dishes is sometimes more acceptable than straight energy supplements which are regarded as a medicine.

The psychological effect of prescribing a diet must be borne in mind, particularly with a child. The mother who says 'I feel anxious every time I give my child his supplement' is articulating a real problem. Any treatment that has to be given two or three times a day over a long period can cause friction between parent and child. For example, a child may refuse to take the prescribed diet supplement and then refuses all the more because mother (who appreciates the importance) is so anxious he takes it. Vomiting or other food refusal can result especially if too many calories are given in supplement form or the child is force fed. One of the disadvantages of being able to concentrate energy is that we can get a large number of calories into a small volume, causing appetite to be suppressed. Remember the groups of rats who were fed oil (i.e. a calorie supplement) or water but allowed free access to rat pellets; both groups of rats chose isocalorific intakes.

Another problem of supplementation was illustrated by a child with gross malnutrition because of gastrointestinal problems resulting from a major gut resection at a few days old. Initial intravenous therapy, then much dietetic work over 10 months to provide high energy intake in concentrated frequent feeds, resulted in a reasonably nourished child, but she continued to demand excess food after her malabsorption state had resolved and became obese. Oversupplementation with carbohydrates may also precipitate diabetes mellitus and this should be watched for, particularly in those patients who have a family history of this disease.

The age and clinical state of a child determines what is absorbed and this is particularly important in infants who develop gastroenteritis. Giving hypertonic solutions orally may worsen the malabsorption and will not improve the child's nutritional state. The osmolarity of the feed will therefore be very important and this is where Caloreen has an advantage over glucose. It is also useful as the carbohydrate in a feed based on puréed chicken for children who have disaccharide intolerance and/or milk protein intolerance and so cannot have milk. The carbohydrate has to be in a simple low-osmolar form and Caloreen is well tolerated.

One of the problems of tube feeds is that of diarrhoea. I am very careful to calculate the child's individual need and watch the concentration not only of carbohydrates but also of protein and fat in the feed. If this is done, diarrhoea rarely occurs even in those suffering from a malabsorption syndrome. Most infants will tolerate between 10 and 12 g carbohydrate per 100 ml and, with normal gut, some infants can tolerate 15 to 20 g of carbohydrate per 100 ml. Obviously the type of carbohydrate is of importance.

Fat is utilised as a good source of energy and provided there is no steatorrhoea infants will tolerate 3 to 4 g of fat per 100 ml and in older children 5 to 10 g of fat per 100 ml provided that the dietary fat intake never exceeds that of carbohydrate.

In the child who is well and able to drink freely one can give concentrated calorie supplements as mousses or milk shakes.

The child needing a very low fat diet for lipid disorders may utilise medium-chain triglyceride fat for energy, but in α-β-lipoproteinaemia even this is not recommended. Carbohydrate supplements will have to supply a very large proportion of energy requirements and Caloreen is particularly useful in this situation.

In the organicacidaemias and Maple Syrup Urine Disease the child requires very severe restrictions in natural protein with or without a supplement of specific amino acids. It is important that he gets adequate carbohydrate for energy requirement for instance by taking a 20% Caloreen solution e.g. 150–200 ml/kg body weight/day in infants.

I have purposely not spoken about renal conditions. The principles that we have already heard also apply in children.

Caloreen is a useful tool to the clinician and paediatric dietitian in the many clinical situations I have described.

FURTHER READING

Francis, D. E. M. (1974). *Diets for Sick Children*. 3rd edn. (Oxford: Blackwell Scientific Publications)

Discussion

Dr Mallick—*I was interested in your comment about dealing with the whole family. How often do you find you have to see out-patients?*

Miss Francis—*The amount of dietetic time that is needed depends on the diagnosis and social economic situation of the family. Telephone contact is a big help in reducing the need for frequent attendance.*

Dr Atherton—*In severe adult coeliac disease it has been shown that there is an active secretory state of the small bowel mucosa. Do you ever find this in children? Before the gluten-free diet has time to improve mucosal function, do you ever have to use parenteral nutrition in childhood coeliac disease?*

Miss Francis—*A number of children who have protracted diarrhoea need initial parenteral nutrition; some of these subsequently turn out to have coeliac disease. However, these days, most coeliac disease diagnosed in infancy and childhood is detected while the patient is in a reasonable clinical condition without such complications as sugar intolerance; these children respond quickly to gluten-free diet alone.*

Dr Sherwood-Jones—*What is a 'balanced' diet?*

Miss Francis—*A 'balanced' diet is hard to define, the best guide is to ensure the presence in the diet of all the essential nutrients in absorbable form and to watch the clinical condition—a child who is growing well, is, to me, taking a balanced diet. If a child is not maintaining* *growth rate, one needs to consider whether or not some essential nutrient is missing, for example a trace mineral, vitamins, or inadequate energy or protein.*

Dr Mallick—*We have heard about renal, liver and intestinal problems. Have the dietitians present found other uses for Caloreen in children?*

Miss Bateman—*Some asthmatic children find taking a lot of solid food difficult and some are very emaciated. We have managed to get their weights up to normal by using Caloreen-supplemented diets.*

Summaries

Caloreen and its Use in Sick Children

D. E. M. Francis

In paediatric dietetics natural foods are encouraged whenever possible. Because of its solubility and lack of taste, Caloreen provides a high energy source which can be added unnoticed to such foods to provide required energy. In small children its low osmolarity is particularly useful for liquid feeds. Many special metabolic problems of paediatrics are a challenge to the dietitian since energy-giving foods may be forbidden. Glucose polymers have found a particular place in these conditions.

Le Caloreen et son Emploi chez l'Enfant Malade

D. E. M. Francis

En diététique pédiatrique on utilise, chaque fois qu'il est possible, des aliments naturels. Par suite de sa solubilité et de son absence de saveur, le Caloreen constitue une source importante d'énergie que l'on peut ajouter de façon inapparente à de tels aliments pour apporter l'énergie requise. Chez le jeune enfant, sa faible osmolarité le rend particulièrement utile avec les aliments liquides. En pédiatrie, le diététicien est confronté à de nombreux problèmes métaboliques particuliers car les aliments énergétiques peuvent être interdits. Les polymères de glucose sont dans ce cas particulièrement utiles.

Caloreen und dessen Verabreichung an kranke Kinder

D. E. M. FRANCIS

In der Pädiatrie wird Naturkost so weit als möglich begünstigt. Da es löslich und ohne Geschmack ist, kann Caloreen als unauffällige hochwertige Energiequelle der Naturkost beigefügt werden. Für Kleinkinder ist seine niedrige Osmolarität besonders günstig bei flüssiger Diät. Viele besondere Stoffwechselprobleme der Pädiatrie stellen den Diätetiker vor eine schwierige Aufgabe, weil energiereiche Kost verboten sein kann. Glukosepolymere sind in solchen Fällen besonders geeignet.

Empleo del Caloreen en Enfermedades Infantiles

D. E. M. FRANCIS

En la dietética pediátrica se aconseja, siempre que sea posible, el empleo de alimentos naturales. Debido a su solubilidad y carencia de sabor, Caloreen provee una fuente de alta energía que puede ser agregada a las comidas sin denotar su presencia, brindando el requerimiento calórico apropiado. En niños pequeños es particularmente útil para la alimentación líquida debido a su baja osmolaridad. Muchos de los problemas metabólicos pediátricos constituyen un desafío para el dietólogo dado que los alimentos que proveen energía pueden estar prohibidos. Los polímeros de la glucosa ocupan un lugar especial en el tratamiento de esos casos.

5

Treatment of Malabsorption in Cystic Fibrosis

H. K. BERRY, M. M. HUNT,
F. W. KELLOGG and R. L. INGBERG

Among the earliest symptoms of cystic fibrosis (CF) are failure to thrive because of lack of enzymes necessary for digestion of food. This feature of the disease, leading to chronic malnutrition, has received relatively little attention until recently. However, as early as the 1940s workers showed that while whole protein produced either no elevation, or a drop in amino nitrogen following ingestion by patients with cystic fibrosis, protein hydrolysates were readily absorbed[1]. These investigators suggested that the digestive deficiencies could be circumvented by using predigested nutrients. Replacement pancreatic enzymes became available about the same time, and little further interest was shown in the use of amino acid hydrolysates for patients with cystic fibrosis. Although supplemental enzymes aid in digestion, it is not possible to completely overcome the malabsorption symptoms even with very high doses.

Interest in predigested nutrients revived with the report by Allan *et al.*[2] on the use of an artificial diet using Caloreen as energy source and Beef Serum Protein Digest as a supplement to ordinary diet and conventional management of patients with cystic fibrosis. They observed increasing rates of weight gain and linear growth, as well as increased energy and activity, improved hair texture and growth and reduction of lung infections.

We had the opportunity to try the supplement of easily absorbed nutrients for cystic fibrosis patients in Cincinnati. Initially the most severely affected patients in the CF clinic were chosen to receive the diet. These were patients who had either failed to gain or who had lost weight during the past 2 years. Extensive laboratory studies were carried out before and at regular intervals following the institution of dietary therapy so that data could be obtained to assess the nutritional status of the children. A group of less severely affected children, not receiving the diet, was studied at the same time for comparison. The diet was supervised by a nutritionist at all times. No attempt was made to restrict other foods in the diet. Furnished in the form of a dry powder, the diet consisted of Beef Serum Protein Digest and Caloreen. Of the recommended daily caloric intake 50–80% was furnished by the mixture. The powder was mixed with fruit juice, soup or other liquids. Fat supplement consisted of medium-chain triglycerides containing 2.5% safflower oil. Both water-soluble and fat-soluble vitamins were given. Our results were reported in 1975[3]. Patients who received the diet showed significant gain in weight, and significant increase in clinical score. We noted that weight and clinical score showed a significant positive correlation. Weight was also positively correlated with albumin, urea, and cholesterol levels in serum. These measurements were followed in both experimental and control patients. Serum albumin concentrations rose in patients taking the supplement, while albumin concentrations remained unchanged in control patients. Urea nitrogen concentrations in experimental patients rose following dietary supplementation, while concentrations in control patients decreased slightly. Cholesterol concentrations were low in both groups and remained unchanged. Our initial objective of improving the nutrition of children with CF was achieved.

Of the 15 patients in the initial group of patients, two have died; three patients in the control group died during the first year of the study. The supplement was discontinued in all but five of these older patients in 1974 because of our lack of funds for support. Growth progress continues to be satisfactory in the five patients who remain on the diet. One girl is in the second year of college and has recently married. She loses weight promptly when she stops taking the diet. In spite of the monotony of the diet, the general improvement in wellbeing experienced by the patients motivates them to continue it.

However, it is not likely that advanced pulmonary changes in patients with severe manifestations of the disease are reversible. Based on the experience with phenylketonuria and other metabolic disorders[4], it seems reasonable that if the treatment is to be effective, it must be begun early in life before chronic malnutrition and chronic infection begin. I report on the current state of four infants in whom the diagnosis of CF was made before 6 months of age.

Following the diagnosis of CF, conventional therapy was begun for each child, consisting of replacement of pancreatic enzymes, physiotherapy, inhalation therapy, postural drainage, antibiotics for acute episodes of infection, and mist tent therapy. At ages ranging from 6 months to 1 year a formula consisting of the beef serum protein hydrolysate, Caloreen and a 3 : 1 mixture of MCT and safflower oils was substituted for the whole protein-containing formula. The formula

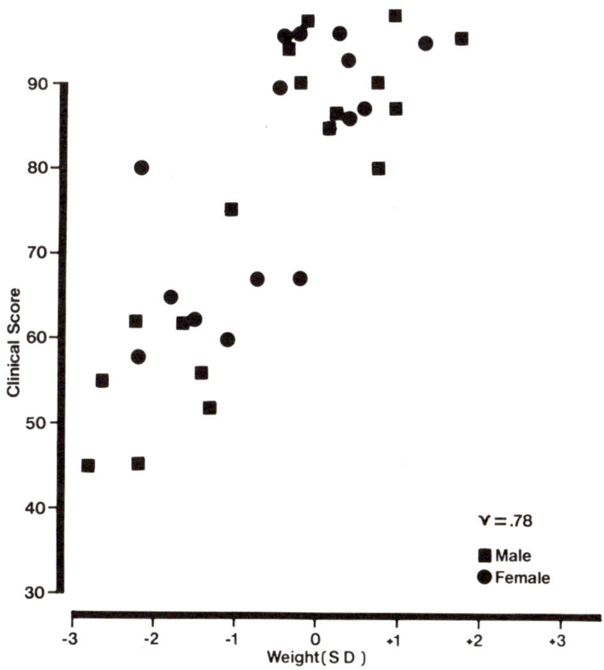

Figure 1 Correlation between the weight (expressed as standard deviations from the mean) and clinical score of patients with cystic fibrosis

was prescribed so that it provided 100% of the recommended daily allowance for protein and 50–80% of the recommended daily allowance for calories—55% of the total calories were in the form of Caloreen. Other foods were added at appropriate ages. No attempt was made to limit intake of other foods. Careful records were kept of intake of the supplement and of other foods in the diet.

None of the children have required hospitalisation during the subsequent 4-year period for CF-related problems.

We had previously demonstrated in the older children a significant correlation between weight and clinical status of children with CF (Figure 1). Weight progress of the four children are shown in Figures 2 to 5. Weights measured at intervals of 6 weeks to 3 months were converted to standard scores based on mean weights of midwestern white children[5]. Mean dietary intakes for 3-month periods were calculated in terms of percent of the recommended daily allowance

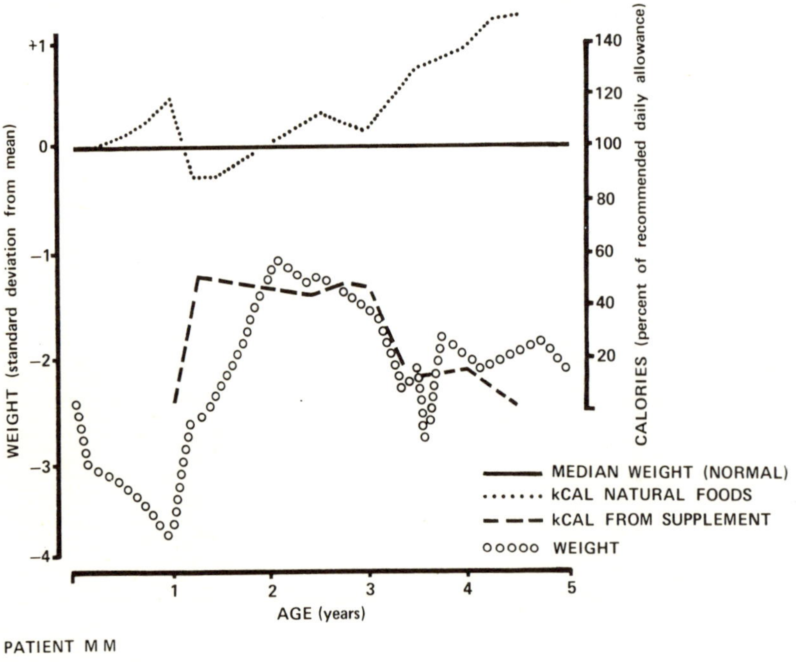

Figure 2 Weight progress of M. M. during first 5 years of life related to intake of calories from supplement of easily absorbed nutrients and from other foods

provided by the nutritional supplement and by other foods in the diet. Data for calories are shown.

Case histories

(1) M. M. was born prematurely with meconium ileus (Figure 2). Sweat chloride test was positive at 6 weeks of age. Replacement pancreatic enzyme therapy as well as other standard therapeutic measures were begun at that time. The formula was changed to Portagen*. He failed to gain weight in spite of the treatment. He was

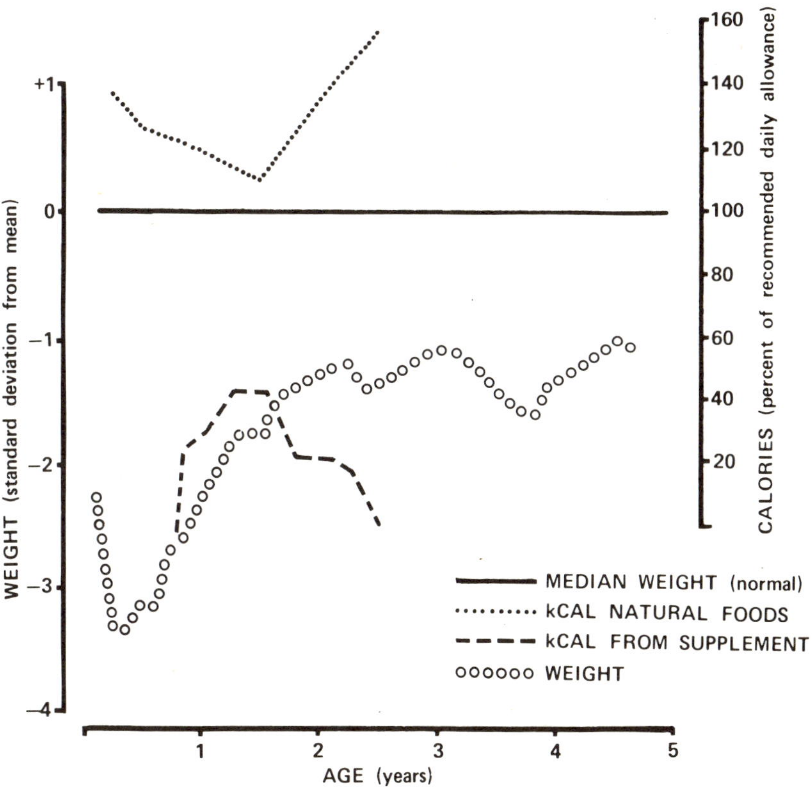

PATIENT L C

Figure 3 Weight progress of L. C. related to intake of calories from supplement and from other foods

* Mead Johnson Company

hospitalised several times during the first year for acute infections and showed progressive lung changes on X-ray. At 1 year the formula of predigested nutrients was substituted for Portagen. He gained weight rapidly, though he remained underweight for age. When he was 3 years old the supplement was stopped because of family difficulties. The child's rate of gain in weight has declined to approximately 2 SD below the mean. The chest X-ray is normal, and there are no changes suggestive of cystic fibrosis.

(2) L.C. was found to have CF at 6 weeks of age when she was hospitalised for failure to thrive and recurrent bronchitis (Figure 3). Conventional CF treatment was begun, including pancreatic enzymes and Portagen. She failed to gain weight. The formula of predigested nutrients was begun at 6 months of age. Although the baby did not take all the prescribed amount of formula, she gained weight rapidly.

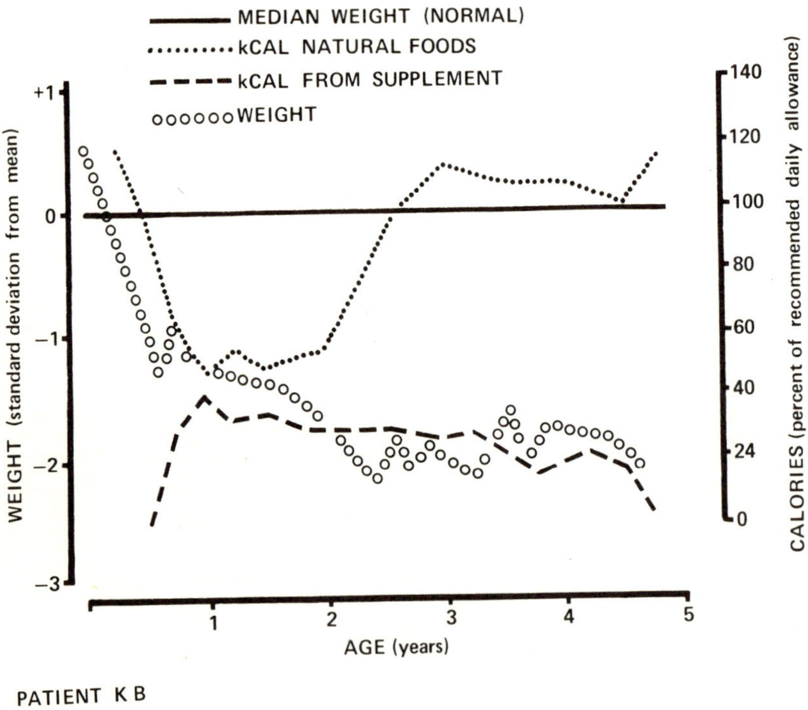

Figure 4 Weight progress of K. B. related to intake of calories from supplement and from other foods

Marital conflict between the parents led to discontinuation of the special diet at age 2½ years. Rate of gain in weight slowed and stabilised at about 1 SD below the mean in spite of continued intake of calories ranging from 160 to 200% of the recommended daily allowance.

(3) K. B. was found to have CF at 7 months of age (Figure 4). He gained weight initially on a formula of Similac* once pancreatic enzymes were begun. The special dietary supplement was started when he was 9 months old. He always took less than the recommended amount. The rate of gain in weight continued to decline. He lost weight during the 6 months after age 4½ when he refused to take the formula, in spite of a caloric intake from other foods 20% above that recommended.

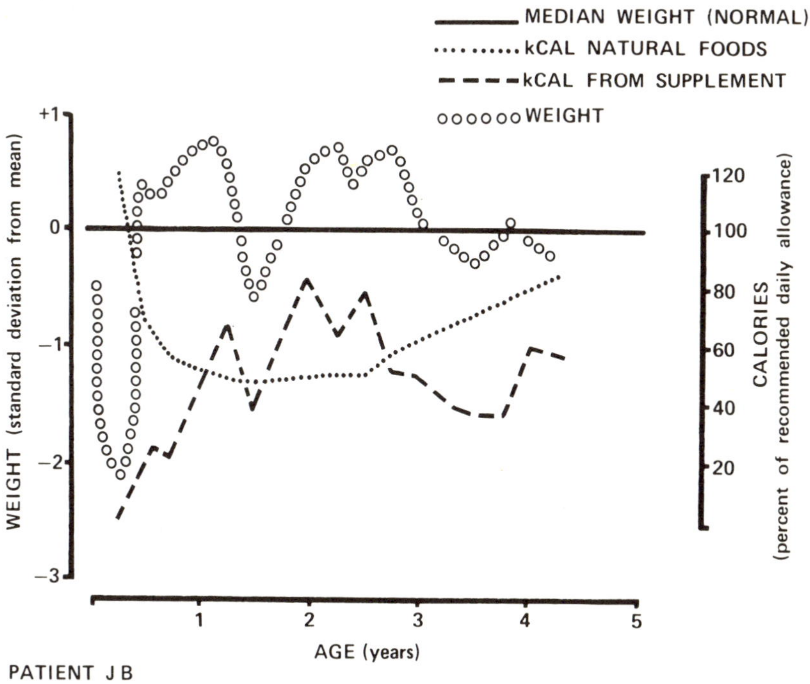

Figure 5 Weight progress of J. B. related to intake of calories from supplement and from other foods

* Ross Company

(4) J. B. was hospitalised for failure to thrive and pneumonia at 3 months of age and a diagnosis of CF was made (Figure 5). She was placed on the formula of predigested nutrients a month later. Weight gain subsequently was rapid. At 18 months of age the amount of formula taken decreased with an accompanying sharp decline in rate of weight gain. After age 2½ when parents were asked to help to pay for the diet, the intake of formula dropped. Rate of gain in weight also declined in spite of the fact that caloric intake remained relatively constant.

Caloric intakes from foods other than the special formula ranging from 20 to 60% above the recommended daily allowance were not sufficient to promote weight gain at the same rates as when 50% of the recommended caloric intake came from easily absorbed nutrients, including Caloreen. The changes in rate of gain in weight appeared to be more closely related to intake of calories from the dietary supplement than to calories or protein ingested in the form of natural foods. The use of predigested nutrients, begun early in life, with emphasis on sufficient amounts of easily absorbed calories should be helpful in preventing the malabsorption symptoms commonly found in patients with cystic fibrosis.

Acknowledgement

Supported by Project 427, Maternal and Child Health Services, Mental Health and Health Services Administration, and grants HD00324 and HD05221 from the National Institutes of Health, USPHS and the Ohio Department of Health Unused Consumer Fund.

REFERENCES

1. West, C. D., Wilson, J. L. and Eyles, R. (1946). *Am. J. Dis. Child.,* **72,** 251
2. Allan, J. D., Mason, A. and Moss, A. D. (1973). *Am. J. Dis. Child.,* **126,** 22
3. Berry, H. K., Kellogg, F. W., Hunt, M. M., Ingberg, R. L., Richter, L. and Gutjahr, C. (1975). *Am. J. Dis. Child.,* **129,** 165
4. Berry, H. K. (1976). *Clin. Perinatol.,* **3,** 15
5. Sontag, L. W., Reynolds, E. L. (1945). *J. Pediatr.,* **26,** 327

Discussion

Dr Mallick—*Do you think that the diet should really be continued indefinitely?*

Prof. Berry—*We can only say that in the older children whose eating habits have already been established we have had to use many devices to persuade them that the change is really good for them. If they stop the loss in weight is usually prompt. I was surprised at the data because I had not been conscious of the relation between the dietary intake, growth and weight until I carried out the analysis.*

Mr Milner—*Is there evidence of malabsorption of sugars?*

Prof. Berry—*We did not carry out xylose tolerance tests but we were curious about why ordinary disaccharides are not just as useful. There are many reports of lactose intolerance in CF. We had observed urinary sucrose excretion in many of our patients for whom a high carbohydrate diet is frequently recommended and the most usual carbohydrate to give is sucrose. Because of this observation we selected some of these patients who had sucrosuria and gave them sucrose tolerance tests, and the rise in blood glucose was perfectly normal. However, these tests were all done in the standard manner with an overnight fast and no other food was given with the sucrose. We are proposing a rather interesting experiment of combining sucrose with a protein load to see if one affects the metabolism and absorption of* *the other. Certainly studies to define abnormality in digestive enzymes in CF are likely to help in prescribing the correct diet.*

Mlle. Delhaye—*In your first experiments you measured blood urea, nitrogen, sera albumin and cholesterol. You said that there was no change in the levels for one year on the diet but is the cholesterol level the same in patients and in controls?*

Prof. Berry—*Cholesterol level is low and remains low in the patients as compared to normal children.*

Miss Francis—*Have you considered Caloreen or a similar supplement alone without the full Albumaid/Caloreen type regime?*

Prof. Berry—*We are beginning to look at using, for example, lipids or Caloreen alone as energy source. I suspect that in severely malnourished patients such as we were studying some nitrogen supplement would be absolutely necessary.*

Mr Milner—*If you feed medium-chain triglycerides can you assume that they have an effect?*

Prof. Berry—*The notion is that they do provide increased energy since while ordinary fats are not well digested, medium-chain triglycerides should be absorbed and pass into the portal vein. There are many studies on the metabolism of medium-chain triglycerides but it was not our particular objective to study them; our nutritionist has considerable experience in using the medium-chain triglyceride milk which I described and found that it alone did not promote normal weight gain in infants to whom it was fed. So we felt that giving medium-chain triglycerides alone would not solve that problem of providing energy for our patients.*

Dr Mallick—*I think there is a dietetic point here—the need to provide a palatable preparation.*

Prof. Berry—*I must confess that another reason we felt the infants would do better is that they have no choice but to take what the* *mother gives them and they come to accept it just as the phenylketonuric children accept their rather unpalatable diet. Most older children turn up their noses at these diets so we thought that by starting in infancy we might overcome the inherent unpalatability of our prescription.*

Miss Cootes—*I could perhaps add to Dr Berry's experience on using Gastro-Caloreen alone for CF. I am quite convinced that we do maintain weight in this way and improve it in some cases without added protein supplement, providing the diet is good in other ways, Caloreen and Gastro-Caloreen are very useful in CF patients who are able to tolerate it in greater quantities than other carbohydrates. However, it is very difficult to say that any one item alone is responsible for the dramatic improvement of some of the very ill children in whom we have used the predigested type of formula, together with Caloreen.*

Summaries

Treatment of Malabsorption in Cystic Fibrosis

H. K. Berry, M. M. Hunt, F. W. Kellogg and R. L. Ingberg

Caloric intakes from ordinary foods ranging from 20–60% above the recommended daily allowance were not sufficient to promote weight gain at the same rate as when 50% of the recommended caloric intake came from easily absorbed nutrients, including Caloreen. The changes in rate of gain in weight appeared to be more closely related to intake of calories from the dietary supplement than to calories or protein ingested in the form of natural foods. The use of predigested nutrients, begun early in life, with emphasis on sufficient amounts of easily absorbed calories should be helpful in preventing the malabsorption symptoms commonly found in patients with cystic fibrosis.

Traitement de la Malabsorption dans la Fibrose Kystique

H. K. Berry, M. M. Hunt, F. W. Kellogg et R. L. Ingberg

Des apports caloriques d'aliments ordinaires compris entre 20–60% au-dessus de la quantité journalière permise ne permettent pas d'obtenir une vitesse d'accroissement du poids semblable à celle qu'on observe lorsque 50% de l'apport calorique recommandé proviennent de substances nutritives facilement absorbées, y compris le Caloreen. Les variations de la vitesse d'accroissement du poids semblent plus étroitement liées à l'apport de calories de l'additif diététique qu'aux calories ou aux protéines ingérées sous forme d'aliments naturels.

L'utilisation de substances nutritives prédigérées commence de bonne heure et on souligne que des quantités suffisantes de calories facilement absorbées sont utiles pour empêcher les symptômes de malabsorption couramment observés chez les patients atteints de fibrose kystique.

Behandlung der Malabsorption bei Mukoviszidose

H. K. BERRY, M. M. HUNT, F. W. KELLOGG und R. L. INGBERG

Kalorienzufuhr aus den üblichen Nahrungsmitteln, die im Bereich von 20–60% über dem empfohlenen Tagesbedarf lag, war nicht ausreichend, eine Gewichtszunahme im gleichen Maß zu fördern, wie bei einer 50%-igen Deckung der empfohlenen Kalorienzufuhr aus leicht resorbierbaren Nährstoffen, einschließlich 'Caloreen'. Die Änderungen im Grad der Gewichtszunahme schienen eher mit der Kalorienaufnahme aus der Diätergänzung in Beziehung zu stehen als mit den Kalorien oder dem Eiweiß aus den verzehrten natürlichen Nahrungsmitteln. Die Verwendung vorverdauter Nährstoffe, wie es im frühen Lebensalter der Fall ist, unter besonderer Berücksichtigung ausreichender Mengen leicht resorbierbarer Kalorien sollte eine Hilfe sein, die bei Patienten mit Mukoviszidose übliche Malabsorptionssymptome zu verhindern.

Tratamiento de la Malabsorción en la Fibrosis Quística

H. K. BERRY, M. M. HUNT, F. W. KELLOGG y R. L. INGBERG

La ingestión de alimentos ordinarios cuyo valor calórico superaba en un 20–60% el requerimiento diario indicado no fue suficiente para promover un aumento de peso de igual magnitud al obtenido cuando un 50% de la ingestión calórica recomendada era derivada de nutrientes fácilmente absorbibles, incluyendo el Caloreen. Los incrementos de peso parecían guardar una relación más estrecha con la ingestión de calorías provenientes del suplemento dietético que con las calorías y proteínas ingeridas en forma de alimentos naturales. El uso de nutrientes pre-digeridos, desde época temprana, contribuiría a la prevención de los síntomas de malabsorción hallados comúnmente en pacientes con fibrosis quística, siempre que se tuviera la precaución de administrar cantidades suficientes de calorías de fácil absorción.

6

Nasogastric Tube Feeding with Caloreen

S. P. ALLISON and A. M. J. WOOLFSON

In this paper, we wish to describe our studies on tube feeding. We have used the glucose polymer Caloreen as the principle calorie source. This follows from the work of Berlyne and his colleagues[1] in the treatment of renal failure. Caloreen has particular advantages as a calorie source. Being a glucose polymer of 5 molecules average chain length, it has only one-fifth the osmolality of glucose in solution and can therefore be given in much larger amounts than glucose without causing diarrhoea. In most patients the whole calorie requirement can be met from this source alone. As mentioned previously glucose is more protein sparing than fat in hypercatabolic patients and since Caloreen is absorbed as glucose it may be expected to be similarly effective. On a purely practical level, an aqueous solution of Caloreen lends itself more readily than fat to administration by continuous drip.

The standard hospital 'nutritional' system consists of a nasogastric tube which is aspirated to keep the stomach empty while normal saline is given intravenously[1]. This does not matter for 3 or 4 days if the patient is then able to take food by mouth. However, when the inability to take food by mouth persists the patient has to be fed in other ways. If the gastrointestinal route cannot be used then intravenous feeding must be employed. However there are very many patients with a normally functioning digestive system who are unable to swallow for one reason or another; in these we employ a nasogastric tube feed. We have considered the problems of tube feeding under the following headings:

Administration

At one time tube feeds were made up in the diet kitchen from various ingredients, put in a jug and sent to the ward where they were kept in the ward refrigerator. If the nurses were not too busy then some of this mixture would be syringed periodically down the nasogastric tube to provide nutrition. The amount of food given in this way was not predictable, and this bolus method tended to produce diarrhoea as the stomach and intestine were challenged intermittently by the sudden injection of a strongly hyperosmolar suspension.

We now have two standard tube feeds, one based on Caloreen and Complan (a preparation containing intact milk protein) and the other on Caloreen and Albumaid[2] (beef serum hydrolysate). These are prescribed by the doctor or dietitian as seems suitable for the individual patient. The feed is easily made up in batches by the diet kitchen cook into 1-litre Winchester bottles and sent to the ward. Using an adapted bladder washout set with the screw connection which fits standard Winchester bottles, the feed is dripped into the stomach via a small Ryle's tube over 24 hours. A work-study[2] showed that our new system saves 1 hour of nursing time per patient per day as compared with the old method of syringing food down the Ryle's tube. If a 3-litre feed is being given then a nurse has only to change the feed drip bottle every 8 hours and check the drip rate periodically. The method of continuous administration also causes much less diarrhoea compared with the bolus method and is more precise since it is easy to determine exactly how much has been given during 24 hours. As the content of the tube feed is known accurately, careful balance of water, electrolytes, nitrogen and calories can be maintained.

Nutritional adequacy

The content of the two tube feeds is shown in the accompanying tables (Tables 1 and 2). The amounts of vitamins and minerals con-

Table 1 Recommended tube feeds. *Caloreen/Complan.* For ordinary maintenance feeding on general wards. Nitrogen is given as whole protein (Complan)

Nutrient	*Weight* (g)	*Non-protein energy* (kcal)	(mJ)	*Protein* (g)	*Nitrogen* (g)	*Na* (mM)	*K* (mM)
Caloreen (glucose polymer)	250	1000	4.2	—	—	—	—
Complan (milk protein)	300	1070	4.5	60	9.6	46	53
Total	550	2070	8.7	60	9.6	46	53

The feed is made up to 3 litres with water. More sodium or potassium may be added as clinically indicated. Hypercatabolic patients (e.g. burns) require both higher energy and nitrogen intakes. Provided the same proportions of energy to grams of nitrogen are observed, the amounts of Caloreen and Complan may be increased to the limit of tolerance of the patient. There is sufficient fat in Complan to prevent fatty acid deficiency. The vitamin, mineral and trace element content are probably sufficient for most patients. Daily cost—£0.59

form to standard recommendations. Each feed gives approximately 200 kilocalories/g of nitrogen. As long as this ratio is maintained the calorie and nitrogen content of each feed can be increased or decreased to suit the particular situation. In some cases double the amounts of Caloreen and Complan have been given and tolerated quite well by the patient. Complan contains sufficient fat to prevent fatty acid deficiency. However, the Caloreen/Albumaid mixture has to be supplemented with egg yolk if it is given for longer than a week. Fatty acid deficiency has been shown to occur in intravenously fed patients after $1\frac{1}{2}$ to 3 months without fat.

Using the Caloreen/Albumaid feed 15 patients were fed and studied for periods between 7 and 41 days. There was no significant weight loss, except in two patients whose initial oedema disappeared with salt restriction and a diuretic. A third patient had an enormous weight gain of 11 kg but this was associated with the unwise administration of intravenous normal saline and not a function of tissue weight change. The Caloreen/Complan tube feed has been used extensively but studied particularly in four patients with head injury who required tube feeding for between 2 and 4 months. Adequate nutrition

Table 2 Recommended tube feeds. *Caloreen/Albumaid*. For special situations where a low electrolyte is required. Nitrogen is given as amino acids derived from the hydrolysis of beef serum (Albumaid)

Nutrient	*Weight* (g)	*Non-protein energy* (kcal)	(mJ)	*Nitrogen* (g)	*Na* (mM)	*K* (mM)
Caloreen (glucose polymer)	600	2400	10	—	—	—
Albumaid (beef-serum hydrolysate)	60	—	—	8	26	3
Mineral mixture (Ca, P, Mg, trace elements)	8	—	—	—	14	17
Ketovite tabs, 2						
Ketovite syrup, 5 ml						
Total		2400	10	8	40	20

If the feeding period is >7 days, essential fatty acids are provided by 1 egg yolk daily. Water is added as appropriate—usually 1500 to 2500 ml. The electrolyte content may be diminished by leaving out the mineral mixture, or increased by addition of sodium or potassium chloride. Daily cost—£2.60 daily

was maintained over this period of time. There was some diffuse atrophy of the legs but apart from this the patients presented a normal and well-nourished appearance. With large amounts of Caloreen, a high blood sugar is sometimes produced in the severely ill patient. We therefore test the urine regularly with Clinitest and maintain a diabetic type urine chart. In the presence of glycosuria greater than 0.25% we administer insulin on a sliding scale. The insulin requirement disappears when the acute phase of the illness has passed. In some patients a crude protein balance was assessed using urea production rate data and in all these patients approximate balance was maintained. In most patients there may be some advantage in presenting a diet of protein hydrolysate as opposed to giving one of whole protein[3]. However where digestive mechanisms are impaired, the use of an amino acid preparation such as Albumaid is preferable, and easy to administer by tube feed.

Water and electrolytes

The main advantage of the Caloreen/Albumaid diet over Caloreen/Complan has been in cases where a low electrolyte content is required (see Tables 1 and 2).

One of the metabolic effects of severe illness is to induce an inability to excrete a large water or sodium load. This may be exacerbated by renal underperfusion from blood volume deficiency or heart failure, by acute renal failure or by positive pressure ventilation. For this reason it is desirable to have basic tube feeds in which the electrolyte content is low but to which further sodium or potassium can be added as the clinical situation demands. It is always possible to add but never to subtract electrolytes from such a preparation. The omission of the mineral mixture from the Caloreen/Albumaid tube feed gives a very low sodium feed particularly suitable for use in acute renal failure. The potassium content can be similarly manipulated. It may be desirable to keep this low in some patients with acute renal failure, whereas in other patients it may be desirable to achieve a high potassium input. In all cases the required electrolyte content is prescribed by the medical staff and the appropriate formula made up in the diet kitchen. By monitoring weight, plasma electrolyte values and, in some cases, by the use of cumulative balance charts, all our patients have been maintained in satisfactory water and electrolyte balance without recourse to intravenous supplementation. The one exception to this has already been mentioned (page 43).

Complications

The main complications of tube feeds in the past have been

(a) diarrhoea,
(b) nausea and vomiting with aspiration of the feed,
(c) hyperosmolar states,
(d) complications due to the Ryle's tube itself, e.g. oesophagitis.

Of the 15 patients receiving Caloreen/Albumaid, seven developed diarrhoea to some degree (more than one loose motion daily). Of these, six were on broad spectrum antibiotics. Of the eight who did not develop diarrhoea, only one was receiving broad spectrum antibiotics. In all cases the diarrhoea ceased within a few hours following the administration of codeine phosphate syrup at a dose of 60–120 mg daily. The Caloreen/Complan tube feed produced few problems and again any diarrhoea that did occur responded rapidly to codeine phosphate syrup. In no case was it necessary to diminish or withdraw the tube feed. The rate of administration of the feed proved an important factor; inadvertent delivery of a large bolus produced diarrhoea which ceased if the rate of administration was reduced. Another factor to which diarrhoea has been ascribed is the high osmolality of tube feeds. This may be true of those containing glucose. Caloreen, which has one-fifth of the osmolality of glucose, weight for weight, has an advantage in this respect.

Nausea and vomiting were largely obviated by giving 60 ml of water and milk per hour at the beginning of the tube feed and aspirating 4-hourly to ensure that this liquid had passed out of the stomach. Only when we were sure that the stomach was emptying properly did we start the tube feed, at half strength for the first day and then at full strength. A few patients experienced slight nausea in the first two days; this was overcome by the administration of Maxolon syrup 10 mg t.d.s.

No cases of serum hyperosmolality were seen. This condition is usually due to one of three factors.

1. If an inadequate number of calories per g of nitrogen are administered in the feed there is a high urea production rate and hence a urea-induced osmotic diuresis; in turn, this may lead to plasma hyperosmolality.
2. An inadequate water intake will have the effect of increasing plasma osmolality even in the presence of a normal urea production rate. The $2\frac{1}{2}$–3-litre water content of our feed was sufficient to prevent this.
3. Another factor is the high blood glucose which may follow a high carbohydrate intake. This was always detected in our patients by urine testing and estimation of blood sugar. Appropriate amounts of insulin were then given to correct hyperglycaemia.

No cases of oesophagitis or dysphagia were detected subsequent to a period of tube feeding. This may be partly due to the constant drip method of administration which continually dilutes stomach acid and yet keeps the volume of gastric juice down to a point where reflux is insignificant.

Cost

We have published data on comparative costs of our tube feeds and other commercially available prepacked feeds[4,5]. The Caloreen/Complan feed costs approximately 60 pence daily and the Caloreen/Albumaid feed about £2.50 daily. Prepacked, so-called 'elemental' diets cost between £5.00 and £12 daily. Such expense cannot be justified when highly effective, well-tried alternatives are available at a fraction of the cost.

Conclusions

We have found two standard tube feeds, one based on Caloreen/Albumaid and another on Caloreen/Complan, to be entirely satisfactory in a large number of patients treated for up to 4 months. The basic feed was modified by the addition of potassium or sodium, as necessary and in a few hypercatabolic patients the total calorie and nitrogen content of the feed was increased. The introduction of a standard procedure for prescribing, preparing and administering a tube feed has increased the efficiency of delivery to the ward, diminished nursing time, increased the accuracy of administration and diminished complications. The occasional incidence of diarrhoea was usually associated with an unduly rapid rate of administration of the feed, or with the prescription of broad spectrum antibiotics. It was easily controlled by adding codeine phosphate syrup to the tube feed.

REFERENCES

1. Berlyne, G. M., Booth, E. M., Brevis, R. A. L., Mallick, N. P. and Simons, P. J. (1969). *Lancet*, **i,** 689

2. Woolfson, A. M. J., Saour, J. N., Ricketts, C. R., Pollard, B. J., Hardy, S. M. and Allison, S. P. (1976). *Postgrad. Med. J.*, **52**, 678
3. Woolfson, A. M. J., Knapp, M. S. and Allison, S. P. (1976). *Lancet*, **i**, 365
4. Allison, S. P. and Woolfson, A. M. J. (1975). *Lancet*, **ii**, 507
5. Woolfson, A. M. J., Pollard, B., Hardy, S. M. and Allison, S. P. (1975). *Lancet*, **ii**, 1157

Discussion

Dr Mallick—*What nitrogen source did you use for parenteral feeding?*

Dr Allison—*A preparation of synthetic* L-*amino acids*

Dr Mallick—*And did you give insulin by infusion or intermittently?*

Dr Allison—*We prefer to give insulin continuously by syringe pump.*

Miss Booth—*If you have a patient who has been on parenteral nutrition and is changed over to tube feeding, do you have any problems?*

Miss Francis—*In paediatrics, the best way to cope with this is by a step-by-step regime decreasing the intravenous nutrition and increasing the oral nutrition or nasogastric tube feed regime slowly over perhaps 10 days.*

Dr Allison—*This problem of dovetailing one kind of nutrition into another is important. One must avoid a gap in which actual intake is not carefully monitored and so nutrition is inadequate. One must ensure that the patient is really taking enough by mouth.*

Dr Berry—*Are the lecithins in the egg yolk adequate to maintain a stable emulsion when you add this to your nutritional prescription in longer-term patients?*

Dr Allison—*Egg yolk will mix well since it is a stable emulsion itself.*

Summaries

Nasogastric Tube Feeding with Caloreen

S. P. ALLISON and A. M. J. WOOLFSON

Caloreen has been used as the principal calorie source in a regimen of nasogastric tube feeding in which either the low-electrolyte protein supplement Albumaid, or Complan is given to provide nitrogen. The low osmolarity of Caloreen solution has made it possible for adequate calories to be given without causing gastrointestinal upset. Extensive experience with these tube feeds has shown that they provide effective nutrition over many weeks.

Alimentation par le Caloreen avec une Sonde Nasale Gastrique

S. P. ALLISON et A. M. J. WOOLFSON

On a utilisé le Caloreen comme source principale de calories, dans une alimentation par sonde nasale gastrique, en apportant de l'azote avec l'additif protéique à faible teneur en électrolytes qu'est l'Albumaid ou le Complan. La faible osmolarité de la solution de Caloreen permet un apport convenable de calories sans provoquer de troubles gastro-intestinaux. Une expérience importante de l'administration à la sonde de tels aliments a montré qu'ils permettent une nutrition efficace pendant de nombreuses semaines.

Nasen/Magensondenernährung mit 'Caloreen'

S. P. ALLISON und A. M. J. WOOLFSON

'Caloreen' ist als Hauptkalorienträger bei einer Diät mittels Nasen/Magensondenernährung verwendet worden, wobei entweder die Eiweißergänzung mit niedrigem Elektrolytgehalt 'Albumaid' oder 'Complan' verabreicht wurde, um Stickstoff zu liefern. Die niedrige Osmolarität der 'Caloreen'-Lösung ermöglicht es, ohne gastro-intestinale Beschwerden ausreichend Kalorien anzubieten. Die umfangreiche Erfahrung mit dieser Sondenernährung hat ergeben, daß sie über viele Wochen eine ausreichende Nahrungsversorgung sicherstellt.

Alimentación con Caloreen por Sonda Naso Gástrica

S. P. ALLISON and y A. M. J. WOOLFSON

Caloreen ha sido empleado como la fuente calórica principal en un régimen de alimentación por sonda naso-gástrica, con el agregado del suplemento proteico de bajo contenido electrolítico Albumaid, o de Complan, para proveer nitrógeno. La baja osmolaridad de la solución de Caloreen ha permitido la administración de calorías adecuadas sin provocar trastornos gastro-intestinales. La extensa experiencia con este tipo de alimentación por sondas ha demonstrado que mantiene una nutrición efectiva por muchas semanas.

7

The Significance of Gluconeogenesis in Starved and ill Patients

S. P. ALLISON and A. M. J. WOOLFSON

Before considering the feeding of patients with severe illness, it is important to have an understanding of the physiological changes which are associated with starvation and illness, for these affect nutritional requirements and determine the design of appropriate diets.

In the fasted state we depend mainly on our fat reserves to provide calories. Our immediate carbohydrate reserves, in the form of glycogen, only total 400 g (1600 kilocalories) and are quickly used up. Although we mobilise our fat reserves to provide calories, a continued supply of glucose is essential for metabolism. Triglyceride, save for its small glycerol moiety, cannot be reconverted to carbohydrate, and essential glucose can only be derived from the protein stores found largely in muscle (gluconeogenesis). Thus the muscle mass has not only a locomotor function, but a storage function analogous and complementary to that of adipose tissue. In starvation[1], the rate of breakdown of protein stores proceeds in a carefully controlled manner. The central nervous system, for example, may adapt by metabolising ketones as part of its energy source, thus limiting the requirement for new glucose. In contrast, as Cuthbertson[2] originally demonstrated, injury is associated with a high rate of protein catabolism which can be diminished but not abolished by feeding. This phenomenon is proportional to the severity of injury and is accompanied by increased calorie requirement, and hence increased oxygen consumption. Kinney[3] showed that 80–90% of this increased energy requirement after injury is met from fat stores. He therefore concluded that the main result of the increased protein catabolism was to provide essential carbohydrate constituents for intermediary metabolism and so for gluconeogenesis. He further showed that raising the blood glucose to a level which would normally switch off gluconeogenesis by the liver in the starved state, failed to do so after injury. The metabolic response to injury may therefore be distinguished from starvation

(a) by a higher metabolic rate,
(b) by a greatly increased and relatively unrestrained rate of gluconeogenesis from protein.

These changes result in a wasting away of adipose tissue and a gradual dissolution of the muscle mass. In starvation, 100% mortality is reached when 40–50% of the body weight has been lost. After injury, when weight loss is more rapid, death may ensue when only 25–30% of the body weight has been lost. It is a matter of the highest importance, therefore, to provide adequate nutrition to our patients.

We have been interested in the way that the metabolic response to injury is mediated. Our own work[4,5] has shown that immediately after injury, insulin response to glucose is impaired and thereafter, although insulin levels are normal or even elevated, there is resistance to its action, perhaps mediated by the elevated levels of catecholamines[6], glucagon[7] and cortisol[8] which occur. This alteration in the balance between insulin and its hormonal antagonists is compatible with the metabolic changes that I have outlined. Insulin enhances glucose oxidation and glycogen formation whereas catecholamines and glucagon cause breakdown of glycogen. Synthesis of triglyceride by adipose tissue cells is insulin-dependent whereas the reverse process of lipolysis is under the influence of catecholamines and glucagon. The uptake of amino acids into muscle cells and their incorporation into protein is again insulin-dependent and the reverse process of muscle breakdown is enhanced by cortisol. The breakdown of amino acids within the liver and the formation of urea and new glucose is resisted by insulin and enhanced by cortisol and by glucagon. From these observations we have drawn several conclusions:

1. The urea production rate will provide a very useful index of the rate of muscle catabolism in seriously ill patients.
2. Carbohydrate may be more protein sparing than fat in the catabolic situation.
3. The therapeutic use of insulin may prevent the excessive degree of protein catabolism which occurs in all severely injured patients and is especially seen where there is extensive tissue loss as in severe burns[9].

We have just completed studies which confirm our hypotheses and show that in the non-catabolic starved patient a standard feed of 10 g of nitrogen covered by 200 kilocalories/g in the form of fat and carbohydrate is entirely satisfactory in limiting protein catabolism to a minimum. The provision of a further proportion of a carbohydrate or the use of insulin makes no difference to this rate. However, in more catabolic patients where the urea production rate rises above 12–15 g/day, we have shown that (a) glucose is more protein sparing than fat and (b) the provision of exogenous insulin further decreases the catabolic rate even when large amounts of carbohydrate are being administered.

These observations have important implications for intravenous feeding and to a lesser extent for nasogastric feeding which is the subject of our next paper.

REFERENCES

1. Cahill, G. F. and Owen, O. E. (1968). *Carbohydrate Metabolism and its Disorders*, Vol. 1., F. Dickens, P. J. Randle and W. J. Whelan, eds. (London: Academic Press)
2. Cuthbertson, D. P. (1930). *Biochem. J.*, **24,** 1244
3. Kinney, J. M., Duke, J. H., Long, C. L. and Gump, F. E. (1970). *J. Clin. Pathol.,* **23,** suppl. 4, 65
4. Allison, S. P., Hinton, P. and Chamberlain, M. J. (1968). *Lancet,* **ii,** 1113

5. Allison, S. P. (1974). *Parenteral Nutrition in Acute Metabolic Illness*. H. A. Lee, ed. (London: Academic Press)
6. Birke, G., Duner, H., Liljedahl, S-O, Pernow, B., Plantin, L. O. and Troell, L. (1957). *Acta Chir. Scand.*, **114,** 87
7. Wilmore, D. W., Lindsay, C. A., Moylan, J. A., Faloona, G. R., Pruitt, B. A. and Unger, R. H. (1974). *Lancet*, **i,** 73
8. Cope, O., Nathanson, I. T., Rourke, G. M. and Wilson, H. (1943). *Ann. Surg.,* **117,** 937
9. Hinton, P., Allison, S. P., Littlejohn, S. and Lloyd, J. (1971). *Lancet*, **i,** 767
10. Woolfson, A. M. J., Heatley, R. V. and Allison, S. P. (unpublished)

Discussion

Dr Ricketts—*In studying urea production did you use 3-day observation periods and average the results?*

Dr Allison—*No, there were a dozen patients studied in all, each one was studied for at least 6 days. At least one cross-over was allowed; we paired each day with the corresponding one in the cross-over period.*

Prof. Berlyne—*You use urinary urea excretion as your index of urea production rate?*

Dr Allison—*Yes.*

Dr Mallick—*Did you relate urea production to body weight?*

Dr Allison—*No, because we used each patient as his or her own control.*

Prof. Berlyne—*Were your patients receiving antibiotics, because these can affect urea production?*

Dr Allison—*Either a patient was on antibiotics or steroids throughout the whole period of study or they were not on these drugs at all.*

Summaries

The Significance of Gluconeogenesis in Starved and Ill Patients

S. P. Allison and A. M. J. Woolfson

Glucose is essential for metabolism and, if it is not provided as nutrition, protein is catabolised to glucose in the liver. If adequate glucose is given to ill patients it can be shown that protein catabolism as measured by daily urea excretion is much reduced, a particularly important point in patients who require intensive care.

L'Importance de la Néoglucogénèse chez les Patients Malades et à la Diète Absolue

S. P. Allison et A. M. J. Woolfson

Le glucose est essentiel au métabolisme et s'il n'est pas apporté par la nourriture, le foie catabolise les protéines en glucose. Si l'on apporte une quantité convenable de glucose aux malades, on constate que le catabolisme protéique, mesuré par l'excrétion journalière d'urée, est considérablement réduit, ce qui est particulièrement important chez les patients nécessitant des soins intensifs.

Die Bedeutung der Glukoneogenese bei Unterernährten und Kranken

S. P. Allison und A. M. J. Woolfson

Glukose ist für den Stoffwechsel unentbehrlich, und wenn sie nicht mit der Ernährung zugeführt wird, wird in der Leber Eiweiß zu Glukose abgebaut. Wenn an Kranke ausreichend Glukose verab-

reicht wird, kann man feststellen, daß der durch die tägliche Harnstoffausscheidung gemessene Eiweißabbau stark vermindert ist, ein besonders wichtiger Punkt bei Patienten, die unter Intensivpflege stehen.

La Importancia de la Gluconeogénesis en Pacientes Emaciados y Muy Enfermos

S. P. ALLISON y A. M. J. WOOLFSON

La glucosa es esencial para el metabolismo, y si no es provista como alimento, las proteínas son catabolizadas a glucosa en el hígado. Si se administra glucosa en forma adecuada a pacientes graves puede demostrarse una reducción marcada del catabolismo proteico, medido por la excreción diaria de urea; este punto es particularmente importante en pacientes que requieren cuidado intensivo.

8

Planning Nutrition in Acute Illnesses

D. B. McWILLIAM, M. ROBINSON and E. SHERWOOD JONES

This paper summarises methods of providing nutrition in the wards and intensive care unit of a district general hospital and is based on experience over the past 18 years. The schemes are therefore well tried and easy to organise. Our observations refer only to adults, and nutrition of the patient with advanced and incurable disease is not considered.

The overall objective is to prevent malnutrition during acute illnesses or following injury. In this context 'malnutrition' means a failure to provide nutrition appropriate to the illness. Nutritional requirements can be simply assessed in terms of energy and nitrogen; the term 'appropriate' means that the nutrition provided maintains an approximate state of metabolic balance. The required intake of energy and nitrogen can be determined by carrying out a metabolic balance on each patient or by using information obtained from other patients with similar illnesses or injuries.

Malnutrition in the hospital in-patient is common and the causes are varied; examples are shown in Table 1. In the first group there is normal gastrointestinal function but the factors listed prevent the patient consuming food supplied in conventional form. In this group failure of intake may go unnoticed or be disregarded owing to the prevalent attitude that since the patient is ill he is expected to have a de-

Table 1 Causes of malnutrition in the hospital in-patient

1. With normal gastrointestinal function		
	Anorexia	
	Apathy, depression, confusion, depressed level of consciousness	
	Paralysis, e.g. stroke, myasthenia	
	Dyspnoea, when severe and persistent	
	Ventilator treatment	
	Hypercatabolism: trauma, burns, sepsis	
2. With gastrointestinal 'failure'		
	Obstruction:	Mechanical
		Paralytic
	Inflammation:	Enteritis
		Colitis
	Fistulae	
	Malabsorption syndrome	

creased intake; in this way a state of semi-starvation remains uncorrected. It is essential to remember that ill patients who are dehydrated because they cannot drink are also starved because they cannot eat.

In the second group, nutrition cannot be given via the gastrointestinal tract or it is desirable to avoid it. The disorders in this group provide the major indications for parenteral nutrition, but such patients are a minority of those who require nutritional aid in hospital wards.

The nutrition of the patient in the ward or intensive care unit can be met by one of four schemes of nutrition which are summarised in Table 2, but in addition a clear policy is required. Firstly, when it is not possible to use conventional food (scheme 1) one of the remaining schemes must be quickly adopted; delay can only result in malnutrition. The second principle is that in any individual patient the scheme should be changed as metabolic requirements and gastrointestinal function alter during the course of the illness.

Scheme 1. Conventional food. It is appreciated that hospital food can usually provide sufficient energy and nitrogen for many illnesses and this has been so since the earliest records (see Table 6). However, an adequate intake is rarely possible during acute or severe illness because of anorexia or for the other reason shown in Table 1.

Scheme 2. The simplest methods to aid consumption are to liquidise

Table 2 Schemes for nutrition of adult in-patients

Scheme	*Applications*
1. Conventional food	Convalescent patient Geriatric patient During complex diagnostic investigations
2. Conventional food puréed in a blender	Geriatric patient Drowsy confused patient Tube feeding
3. Liquid diets—consumed orally or by intragastric tube.	Critically ill/injured Coma Ventilator patients
4. Balanced intravenous nutrition	Gastrointestinal 'failure' (see Table 5)

conventional food into a purée, or to provide liquid diets, which are given as beverages or tube feeds. Beverages are drunk by the patient and require palatability to ensure adequate intake. There is a wide range of commercial preparations available for use from simple supplements to full elemental diets. Variety is possible by using liquid diets based on milk protein with added calories in the form of Caloreen (see Table 3 for recipes and costs). Alcohol and chocolate are useful energy sources which can be used to improve palatability.

Scheme 3. Tube feeds are used when the oral route is impractical. This means a nasogastric tube as we never see indications for gastrostomy or jejunostomy in acute illnesses. Here of course palatability is not a factor. John Hunter (1790) was the first to describe intragastric feeding. He used a fresh eel-skin introduced with a whalebone stilette and via a bladder and pipe 'injected jellies, eggs beat up with a little water, sugar and milk or wine'[1]. For more than 10 years we have used as a standard intragastric tube feed (Table 4) a modern version of Hunter's diet. It has been shown to provide satisfactory nitrogen balance in all but the most severe cases of trauma or sepsis associated with a hypercatabolic state[2]. The use of a liquid diet of constant composition was an important step forward in the nutrition of the

Table 3 Recipes and costs

High protein drink (Adapted from Aberdeen School of Dietetics)

500 ml milk
1 packet Carnation Instant Breakfast Food
100 g Gastro/Caloreen
1 egg
15 g Casilan

48 g protein (0.56 pence/g)
4240 kJ (7.19 pence/mJ)

If $\frac{1}{2}$ sachet of Kelloggs Two Shakes is substituted for the Carnation Instant Breakfast Food, this gives:

42 g protein (0.58 pence/g)
4000 kJ (6.13 pence/mJ)

Egg and orange drink (Adapted from the Royal Infirmary, Edinburgh)

Mix 1 egg yolk with 50 ml evaporated milk,
100 ml orange juice and 50 g Caloreen.
Whisk egg white and fold in. Served chilled.

12.7 g protein (1.18 pence/g)
1750 kJ (8.57 pence/mJ)

Tomato soup

100 ml tinned tomato soup
100 ml milk
50 g Caloreen

4.3 g protein (12.4 pence/g)
1400 kJ (8.86 pence/mJ)

Whiston intragastric liquid diet (see Table 4)

1.08 pence/g protein
8 pence/mJ

critically ill because it enabled a metabolic balance to be carried out with minimal facilities[3]. The daily intake of 3 litres is given in aliquots of 125 ml each hour.

Even if the nutritional content of the liquid feed is satisfactory, complications may still occur. These are related to hyperosmolarity, to high lactose content of the feed and to high protein concentration. Hyperosmolarity occurs when the feed is highly concentrated—since sugars are a major component—and results in vomiting and

Table 4 Whiston intragastric liquid diet (1976)

	Formula		
	Complan	150 g	
	Caloreen	75 g	
	Methyl cellulose	3 g	
	Water to make	1 litre	
	Composition		
	Units	1 litre	3 litres
Protein	g (N)	30 (4.8)	90 (14.4)
Fat	g	24	72
Carbohydrate	g	158	474
Na	mmol	22	68
K	mmol	32	98
Ca	mmol	27	81
Mg	mmol	3.5	11
Energy	kJ (kcal)	4057 (966)	12 172 (2898)
Water	ml	835	2610
Osmolarity	mmol/kg	626	626

diarrhoea. It may be avoided by checking the osmolarity of the feed which should be close to 300 mmol/kg initially and then increased as tolerated. The use of glucose polymers such as Caloreen will significantly reduce the osmolarity while providing the same caloric intake as would sugars in the form of glucose or sucrose.

High lactose content may result in diarrhoea as lactose intolerance in adults is relatively common; an incidence of between 25–55% has been quoted[2]. Diarrhoea following tube feeding may also be the result of abrupt change to a low roughage diet and it is for this reason that methyl cellulose is added to our standard intragastric diet (Table 4). Constipation rather than diarrhoea is the more common symptom in patients on this tube feed. A feed of high protein concentration may so increase urea production during its metabolism that there is a progressive rise in the blood urea which should not be mistaken for acute renal failure.

Scheme 4. Parenteral nutrition will be required in patients with gastrointestinal failure and certain other conditions (Table 5). Technically it is more exacting. A central venous catheter is required and strict aseptic and antiseptic techniques are necessary to prevent

Table 5 Indications for intravenous feeding[6]

I. Gastrointestinal 'failure'

Dysphagia
Pyloric stenosis
Intestinal obstruction
Intestinal fistula
Ulcerative colitis
Crohn's colitis
Paralytic ileus (due to peritonitis, abdominal trauma, pancreatitis, head injury, encephalitis, subarachnoid haemorrhage, hypoxia, endotoxaemia)
Acute pancreatitis
Acute cholecystitis

II. Supplement food or intragastric feeding

Hypercatabolic state due to burns, sepsis, multiple injuries

III. Preparation for surgery

All patients with malnutrition

IV. Anorexia nervosa

infection. The range of nutrients available commercially is wide and many symposia and books are devoted entirely to their selection and use. Again we base our methods on a standardised regimen[4] which with minor modifications has proved effective for more than 10 years. It consists of 1.5 litres of Aminosol–Fructose–Ethanol, 1.0 litre of Intra-lipid and 0.5 litre of Vamin fructose or dextrose (Kabi Vitrum products) per day. This provides 11 g of nitrogen and approximately 14 700 kJ (3500 kilocalories per day). In standard form, 6 g of potassium chloride are added daily (2 g to each bottle of Aminosol–Fructose–Ethanol) and one ampoule of Multibionta (Merck), a parenteral multivitamin preparation. It is given at a standard rate of 1 litre 8-hourly. The feeds are infused in pairs—one of Aminosol–Fructose–Ethanol and one of Intra-lipid or Vamin via a three-way tap connected to the central venous line. As gastrointestinal function returns, the parenteral programme may be supplemented with one of the previously mentioned techniques until conventional food can be taken.

One problem with all techniques of nutrition in the acutely ill, is the high incidence of glucose intolerance, a particular problem in the

Table 6 Diet approved by St Bartholomews Hospital, April, 1687

Dyett appointed

Wednesday:	10 oz of bread
	4 oz of cheese
	2 oz of butter
	1 pint of milk pottage
	3 pints of beere

(from Drummond and Wilbraham[7])

Approximate composition of above diet with current recommendations in brackets (sedentary adult—D.H.S.S. 1969)

Energy kJ	9450	(11 340)
Protein g	70	(68)
Fat g	101	
Calcium mg	1224	(500)
Iron mg	13	(10)

This shows the diet to be of reasonable composition apart from the lack of fruit and vegetables.

patient requiring intensive care. Blood glucose should be monitored and glucose intolerance controlled by continuous infusion of insulin given via a clockwork-driven syringe[5].

REFERENCES

1. Jones, E. S. (1969). *Trans. Hunterian Soc.* (*London*), **26,** 75
2. Peaston, M. (1967). *Postgrad. Med. J.* **43,** 317
3. Jones, E. S. and Sechiari, G. P. (1963). *Lancet*, **i,** 19
4. Peaston, M. (1966). *Br. Med. J.* **ii,** 388
5. McWilliam, D. B. (1977). *Clin. Sci. Mol. Med.*, **52,** 25p
6. Jones, E. S. (1977). *General Intensive Care.* (Lancaster: MTP Press Limited)
7. Drummond, J. C. and Wilbraham, A. (1939). *The Englishman's Food* p. 126 (London: Jonathan Cape)

Discussion

Dr Mallick—*Your intravenous amino acid–fat mixture is different from that of Dr Allison. Does each regime have any particular advantage or disadvantage, or do you think they are effectively interchangeable?*

Dr McWilliam—*In many instances I think they would be interchangeable. We occasionally use 50% dextrose solution as a source of calories in addition to, or in place of, the fat solution. The problem of glucose intolerance is greater than with the standard intravenous programme I have described.*

Dr Allison—*I think that the two regimes are complementary. I showed (p. 53) that in the relatively non-catabolic patients who constitute the majority I used exactly the same kind of regime as you have described. We check the urea production rate and if a patient is becoming very catabolic then we switch over to the use of glucose as the calorie source.*

Dr McWilliam—*Yes, we don't claim to correct negative nitrogen balances in hypercatabolic patients with our basic regimen.*

Prof. Berlyne—*Do you get any problems from your 200 g Intra-lipid per day?*

Dr McWilliam—*We do see obvious lipid in the blood when taking blood samples which causes difficulty with electrolyte determinations.*

There is a period of 8 hours in the day when no fat is being administered on our programme so this gives time for adequate clearing of the plasma.

Prof. Berlyne—*Can Dr Allison prevent hypercatabolism and negative nitrogen balance in burns patients?*

Dr Allison—*The hypercatabolism of burns patients is exquisitely sensitive to the environmental temperature; in these cases the thermoneutral zone is around about 33* °C. *Douglas Wilmore has done experiments asking the patients to switch up their temperature control to the level which they find most comfortable, and they appear to switch it to the level at which their hypercatabolism is cut to the minimum. If you nurse patients in a hot environment and so reduce their rate of tissue breakdown to that of the other severely ill patients we are dealing with, then you can switch off their hypercatabolism and put them into positive nitrogen balance.*

Summaries

Planning Nutrition in Acute Illnesses

D. B. McWilliam, M. Robinson and E. Sherwood-Jones

This paper summarises methods of providing nutrition in the wards and intensive care unit of a district general hospital and is based on experience over the past 18 years. The schemes are therefore well tried and easy to organise. Our observations refer only to adults and nutrition of patients with advanced and incurable disease is not considered. Three schemes of nutrition are described and the indication for each is discussed. Caloreen has been used as the main energy source in tube feeds and is a supplement to puréed or conventional food.

La Nutrition Planifiée au Cours des Maladies Aiguës

D. B. McWilliam, M. Robinson et E. Sherwood-Jones

Cet article résume les méthodes de nutrition des services généraux et de l'unité de soins intensifs d'un hôpital général régional et repose sur l'expérience des 18 dernières années. Les protocoles sont donc confirmés et faciles à mettre en pratique. Nos observations ne concernent que les adultes, et la nutrition de patients présentant des maladies avancées et incurables est exclue. Trois protocoles de nutrition sont décrits et leurs indications sont étudiées. On a utlisé le Caloreen comme source énergétique principale pour l'alimentation par sonde et comme additif des aliments réduits en purée ou ordinaires.

Ernährungsplanung bei Akuten Erkrankungen

D. B. McWilliam, M. Robinson und E. Sherwood-Jones

Dieser Vortrag faßt die Methoden zur Erprobung der Ernährung auf den Stationen und in der Intensivpflege-Abteilung eines Bezirkskrankenhauses zusammen und basiert auf den Erfahrungen der vergangenen 18 Jahre. Die Diätpläne sind daher gut erprobt und leicht zu organisieren. Unsere Beobachtungen beziehen sich nur auf Erwachsene, und die Ernährung von Patienten mit fortgeschrittenen und unheilbaren Krankheiten blieb unberücksichtigt. Es werden drei Ernährungspläne beschrieben, und die Indikation für jeden wird diskutiert. 'Caloreen' ist als Hauptenergiequelle bei der Sondenernährung und als Ergänzung zur Schonkost oder üblichen Nahrung eingesetzt worden.

Planificacion de la Alimentación en las Enfermedades Agudas

D. B. McWilliam, M. Robinson y E. Sherwood-Jones

Este trabajo suministra un resumen de los métodos de alimentación en las salas y en la Unidad de Cuidado Intensivo de un hospital general, y se basa en la experiencia de los últimos 18 años. Por ende los esquemas han sido ampliamente utilizados, y son fáciles de organizar. Nuestras observaciones sólo se refieren a adultos, y no se incluyen los pacientes con enfermedades avanzadas e incurables. Se describen tres esquemas de nutrición y se discute la indicación para cada uno. El Caloreen ha sido utilizado como la fuente principal de energía en la alimentación por sonda y es un suplemento de los alimentos en papillas o la comida convencional.

9

Initial Studies into Intravenous Caloreen in Man

S. T. ATHERTON and D. M. WRIGHT

Introduction

Caloreen is used as an oral nutrient in the management of the special dietary problems of patients suffering from renal and hepatic disease and in this context it has found wide acceptance.

In those patients who cannot take oral nutrition and require parenteral feeding, existing carbohydrate regimes remain less than satisfactory for a variety of reasons. Glucose, although readily metabolised, can only be given in strong solution by a central venous catheter. Fructose may accentuate the development of lactic acidaemia in shocked patients, and, like glucose and fructose, sorbitol is readily filtered at the glomerulus and there are heavy urinary losses.

The idea of using Caloreen as an intravenous feeding source is attractive because in an iso-osmolar solution about 1000 calories could be administered in each litre of fluid. We therefore undertook a pilot study to investigate the possibility of using Caloreen as an intravenous nutrient.

Subjects and methods

The questions we attempted to answer were: 'Is Caloreen metabolised to glucose when given intravenously?' 'What effect is seen on the serum insulin?' 'How large are the urinary losses?' Finally—'What are the effects of infusing Caloreen on the peripheral veins?'

The study was performed as a group of three series of experiments. The subjects were all volunteers and all except one was medically qualified. A 30% solution of Caloreen was used and infused into a peripheral arm vein using an Abbott butterfly needle. Blood samples were taken from a vein in the opposite arm kept patent by a continuous slow saline drip. The first few ml of each sample were discarded because of contamination with saline. The samples were analysed for blood glucose, serum amylase and serum insulin. Urine was collected and the amount of carbohydrate present was determined. All the subjects entering the study had normal glucose tolerance tests, fasted overnight and remained in bed for the duration of the experiment.

Results

In the first experiment, three subjects received between 400 and 500 ml of 30% Caloreen by a continuous intravenous infusion over a period of 6 hours, and blood and urine samples were taken at regular intervals.

Table 1 shows the time interval of blood sampling, the glucose levels of the three subjects, and the insulin levels, which were only measured on the first two subjects. There was a steady rise in the blood glucose levels in all subjects, but despite this rise there was no corresponding stimulus to insulin production and the levels of insulin remained basal throughout the experiment.

Table 2 shows the serum amylase levels throughout the experiment. There was a steady fall in the amylase level and this may indicate that amylase is being absorbed on to the dextrin molecule, yet cannot metabolise it completely because of the 1:6 linkages. To ensure that the falling amylase was not due to prolonged starvation the estimation was repeated in the same subjects following a 16-hour fast; the levels throughout this period did not change significantly (Table 3).

None of the subjects developed any sign of inflammation at the infusion site, nor were any other ill effects noticed.

The urinary excretion of total carbohydrate was, however, very

Table 1 **Glucose and insulin levels during an infusion of 400–500 ml Caloreen over a 6-hour period**

	Subjects				
Time (min)	*Blood glucose* (mg/100 ml)			*Serum insulin* (μU/ml)	
	1	2	3	1	2
0	88	106	90	—	9.2
5	92	—	100	18.3	6.3
10	86	—	116	16.2	15.1
15	92	—	116	28.8	10.8
30	90	142	132	14.2	9.5
45	86	140	138	10.8	10.6
60	106	134	140	8.5	12.9
90	114	124	140	11.1	10.2
120	130	124	152	11.0	11.2
150	112	130	152	—	12.5
210	132	136	150	9.8	11.8
270	140	128	144	—	13.9
330	154	140	170	—	11.6

Table 2 **Amylase levels during an infusion of 400–500 ml Caloreen over a 6-hour period**

Time (*min*)	*Serum amylase* (IU/100 ml) *Subjects*	
	1	2
0	195	185
4	180	190
18	185	—
34	—	170
63	180	—
108	150	—
129	—	120
204	—	105
233	130	—
293	130	—
324	—	105
353	130	—

Table 3 Amylase levels during fasting

Time (hr)	*Serum amylase* (IU/l) *Subjects*		
	1	2	3
9	130	145	240
12	125	130	240
15	130	160	235

large, being 56, 44 and 57% of the infused dose in each of the three subjects, so we decided to investigate the behaviour of a smaller infusion of Caloreen over a short period of time and to follow the blood

Table 4 Plasma, glucose and insulin levels and urinary carbohydrate loss during a controlled infusion of 50 ml/h Caloreen for 3 hours)

Time (min)	*Subjects*					
	Blood glucose (mg/100 ml)		*Serum insulin* (μU/ml)		*Urine carbohydrate* (g)	
	1	2	1	2	1	2
0	90	88	11	22		
5	98	96	14	26		
10	103	100	13	21		
15	109	103	16	26		
30	116	103	13	25	0.08	
60	120	114	17	18	0.34	1.64
90	128	122	17	23		5.97
120	132	—	18		1.77	
180	132	124	14	20	9.75	
240	116	112	14	22	4.46	8.35
300	106	100	11	20	0.82	
360	100	96	13	18	0.24	
			Total urine loss (g)		17.46	15.96
			% of dose		39	35

glucose and urinary carbohydrate for a period of 3 hours after the infusion.

Two subjects each received 150 ml of Caloreen by means of a Watson Marlow infusion pump calibrated to administer 50 ml/h; as in the previous experiment there was no evidence of venous damage due to the infusion.

Table 4 demonstrates the blood glucose, serum insulin and urinary carbohydrate levels during and after the infusion. No marked change in the insulin levels occurred. After the infusion had ceased the blood glucose fell rapidly and returned to basal levels within 2 hours. In both subjects urinary loss of carbohydrate continued after stopping the infusion. The total loss of carbohydrate was respectively 39 and 35% of the infused dose.

The lack of any insulin response to the elevated blood glucose levels was puzzling. It may have been due to inhibition of insulin release, to some unknown effect of the infused Caloreen, or our method of measuring blood glucose may have given falsely high readings by measuring the reducing activity of other saccharides in the infusate. A further experiment was designed to test this theory.

Table 5 Plasma, glucose and insulin levels during a controlled infusion of 50 ml/h of Caloreen for 3 hours. Tolbutamide, 1 g, given intravenously after 55 minutes

Time (min)	*Subjects*					
	Blood glucose (mg/100 ml)			*Serum insulin* (μU/ml)		
	1	2	3	1	2	3
0	86	82	94	17	19	17
15	90	92	106	18	24	19
30	98	106	112	17	29	20
55 T	106	100	118	18	23	19
65	92	96	116	48	76	68
75	96	96	120	66	84	52
85	86	80	116	45	60	36
95	62	74	106	29	42	33
100	70	74	106	21	24	29
120	90	94	100	20	27	27
180	100	88	100	16	20	17

T = time of injection of 1 g tolbutamide

falsely high readings by measuring the reducing activity of other saccharides in the infusate. A further experiment was designed to test this theory.

Three subjects were given a controlled infusion of Caloreen by a Watson Marlow infusion pump at a rate of 50 ml/h. Tolbutamide in a dose of 1 g was administered by slow intravenous injection 55 minutes after the commencement of the infusion. Table 5 demonstrates an adequate insulin response to intravenous tolbutamide and subsequent fall in the blood glucose level in the first two subjects, but not in the third. These results seem to indicate that we were in fact measuring true glucose in the first instance.

Conclusions

These studies on the metabolic effects of an infusion of Caloreen in man suggest that there is some degree of breakdown of Caloreen, as evidenced by the rise in blood glucose and the higher proportion of lower molecular weight saccharides in the urine than in the infusate (see p. 80).

Its suitability as a parenteral feeding solution has been examined. Although 30% Caloreen is non-irritant to the vein wall, its metabolic breakdown is very slow and urinary losses are a severe limitation to the intravenous use of the existing preparation. In our study there was an average urinary wastage of 45% of the infused dose; in real terms this means that each litre of 30% Caloreen would only provide 600–700 calories. Even if rapidly utilised this would not be useful as an adjunct to current parenteral feeding regimes where calorie intakes of 3500 per day are required.

The notion of using a glucose polymer as a calorie source is too ingenious to be rejected at this early stage, and studies using solutions of smaller molecules with only 1:4 linkages between glucose molecules should be of value.

Discussion

Dr Mallick—*You pointed out that in percentage terms the loss in the urine is of the order 30–40%. Dr Ricketts who did the appropriate tests will verify, I think, that it was largely the high molecular weight fraction of the compound which was lost.*

Dr Ricketts—*I can only speak qualitatively, but the impression I have is that I am finding the high molecular weight material of Caloreen in the urine soon after infusion and that this is the bulk of the overall urinary loss.*

Dr Mallick—*In other words if one used only a 1:4 linked, short-chain compound, the high wastage rates might be reduced?*

Dr Ricketts—*Yes, I think that suggestion is correct. A 4-glucose unit material only requires to 'see' the enzyme once to be divided into two molecules of maltose, and therefore has a very high probability of feeding someone, but the longer chains need to 'see' the enzyme several times in order to provide maltose, and if there are 1:6 links in the material none of it can be split to disaccharides at all.*

Dr Atherton—*The problem is that if we use a shorter-chain glucose polymer the osmolarity of the solution will rise and the amount we can give in an iso-osmolar solution fall.*

Dr Allison—*A four molecule chain would allow you to give 20% of an isotonic solution, which would be very satisfactory.*

Mr Milner—*I think you should know that we have now confirmed that the Caloreen infusions did contain 25 to 30% of 1:6 linkages. Molecules which are 10 or 12 glucose units long with scattered 1:6 links are indeed unmetabolised by α-amylase.*

Dr Bayliss—*Is the breakdown by amylase or by phosphorylase? Even small oligosaccharides can be broken down by phosphorylase.*

Dr Ricketts—*We haven't thought much about phosphorylase as the enzyme that might break down the saccharides, but we have thought about glucose transferase which is another enzyme capable of transferring glucose units from one chain to another. We found that when maltose was infused some trisaccharide was excreted suggesting that there is an enzyme which is capable of splitting maltose and transferring the glucose onto another molecule. I guess this happens with Caloreen too.*

Dr Bayliss—*Would there be adequate amounts of phosphate present if a phosphorylase was involved; would the addition of phosphates to the Caloreen solution be helpful?*

Dr Ricketts—*I think there isn't enough phosphate present, as you suspect, because one requires 1 molecule of phosphate for each molecule of glucose, and that condition doesn't obtain in the body.*

Dr Davies—*We faced the same difficulty that an appreciable proportion of our Caloreen solution could not be metabolised at the 1:6 links and links at the end of the chain and so on, and probably early after injection or during an infusion most of this was excreted in the urine. We are in a cleft stick; if we have a chain which is long enough to allow us to use fairly concentrated solutions at iso-osmolar concentration then almost certainly, as we have seen between glucose and maltose, the length of time required for metabolism would be longer and during this time the material is circulating and will undergo glomerular filtration. The longer the chains, the slower the breakdown, the more will be lost in the urine. With maltose there was a definite but small urinary loss, more than occurred in the glucose studies. I think we shall have to compromise in the end and accept a degree of loss in the urine.*

Summaries

Initial Studies into Intravenous Caloreen in Man

S. T. ATHERTON and D. M. WRIGHT

Intravenous infusion of Caloreen in volunteers resulted in high urinary excretion of its constituent saccharides and did not produce an insulin response. Tolbutamide infusion was still capable of causing a prompt rise in plasma insulin level suggesting that Caloreen was not preventing insulin release. The results together with analysis of urinary excretion suggest that Caloreen in its present form does not provide sufficient nutrition when given parenterally. It seemed probable that a product in which the saccharides were of small chain length and all linked at the 1:4 position might be effective for this purpose.

Études Initiales portant sur le Caloreen Intraveineux chez l'Homme

S. T. ATHERTON et D. M. WRIGHT

L'injection intraveineuse lente de Caloreen à des volontaires a provoqué une excrétion urinaire importante de ses saccharides constitutifs et n'a pas provoqué de réponse insulinique. L'injection lente de tolbutamide est capable de provoquer un accroissement rapide de la teneur plasmatique en insuline, ce qui suggère que le Caloreen n'empêche pas la libération d'insuline. Les résultats ainsi que l'analyse de l'excrétion urinaire suggèrent que le Caloreen, sous sa forme actuelle, ne permet pas une nutrition suffisante lorsqu'on l'administre par voie parentérale. Il semble probable qu'un produit dont les saccharides auraient une chaîne courte et seraient tous unis en position 1 :4 serait efficace à cet égard.

Primäruntersuchungen mit intravenösem 'Caloreen' am Menschen

S. T. ATHERTON und D. M. WRIGHT

Die intravenöse Infusion von 'Caloreen' an Probanden führte zu einer hohen Harnausscheidung seiner Saccharidbestandteile und ergab keine Insulinreaktion. Die Tolbutamid-Infusion löste noch einen eindeutigen Anstieg des Insulin-Plasmaspiegels aus, woraus hervorgeht, daß 'Caloreen' die Insulinfreisetzung nicht verhindert. Aus den Ergebnissen folgt zusammen mit der Analyse der Harnausscheidung, daß 'Caloreen' in seiner gegenwärtigen Form bei parenteraler Verabreichung keine ausreichende Ernährung bietet. Es könnte sein, daß ein Produkt, bei dem die Saccharide eine kleine Kettenlänge aufweisen und alle in der 1 :4-Stellung verknüpft sind, sich für diesen Zweck verwerten lassen würde.

Estudios Iniciales Sobre el Caloreen Intravenoso en Seres Humanos

S. T. ATHERTON y D. M. WRIGHT

La infusión intravenosa de Caloreen en voluntarios produjo una alta excreción urinaria de sus constituyentes sacáridos y no provocó una respuesta insulínica. La infusión de Tolbutamida mantenía aún su capacidad de causar una rápida elevación en el nivel de insulina plasmático, lo que sugería que el Caloreen no interfería con la liberación de insulina. Los resultados, junto con el análisis de la excreción urinaria, sugieren que el Caloreen en su forma actual no provee una nutrición suficiente cuando es administrado parenteralmente. Parecería probable que un producto en el que los sacáridos tuvieran cadenas cortas, todos con uniones en la posición 1 :4, podría ser efectivo para este propósito.

10

Metabolic Studies Using Infusions of Glucose Polymers and Disaccharides

D. DAVIES and N. P. MALLICK

A carbohydrate can be shown to act in a glucose-sparing capacity if it produces a reversal of the effects of fasting, namely fall in free fatty acid levels, suppression of the secretion of glucagon and growth hormone and a rise in the respiratory quotient (RQ). If glucose is produced from the carbohydrate provided, there will be, in addition, a rise in the level of insulin, with a fall in plasma lactate levels; if fructose is the metabolic source to which the carbohydrate is degraded there is no change in the level of insulin and often a rise in the level of plasma lactate.

We had previously shown that Caloreen taken by mouth is broken down rapidly to glucose and absorbed with the same speed and insulin response as glucose[1]. The intestinal brush border copes easily with 1:4 or 1:6 glucose linkages, breaking down the maltose, maltotriose and limit dextrins released by intraluminal digestion. The availability of a 30% solution of Caloreen for intravenous infusion led to studies to determine whether the material was metabolised and so available as an energy source. However, the problem arose as to whether plasma and hepatic enzymes would be able to accomplish the breakdown of the glucose residue chains. Caloreen is a mixture of oligosaccharides of variable chain length mainly 1:4 linked but with a proportion of 1 : 6 linkages. It is this latter group and those chains in excess of ten glucose residues which seem most likely to be incompletely metabolised. All these compounds undergo glomerular filtration and, in consequence, a high urinary loss of intact fragments would be expected.

When bolus injection studies were carried out using 25 g of the approximately iso-osmolar 30% Caloreen solution, total carbohydrate levels rose and there was a slight rise in blood glucose, perhaps due to some free glucose in the solution. The presence of glucose, maltose and maltotriose in the later urine samples suggested that there was a degree of endogenous breakdown (see p. 8). Insulin levels did not rise and there was no fall in plasma FFA and lactate levels.

When 30% Caloreen was given by continuous infusion, glucose levels and insulin production showed little increase, lactate levels showed a slight downward trend; the RQ showed little increase. When a bolus of glucose was given late in the study, there was a distinct rise in insulin levels and the RQ (Figure 1).

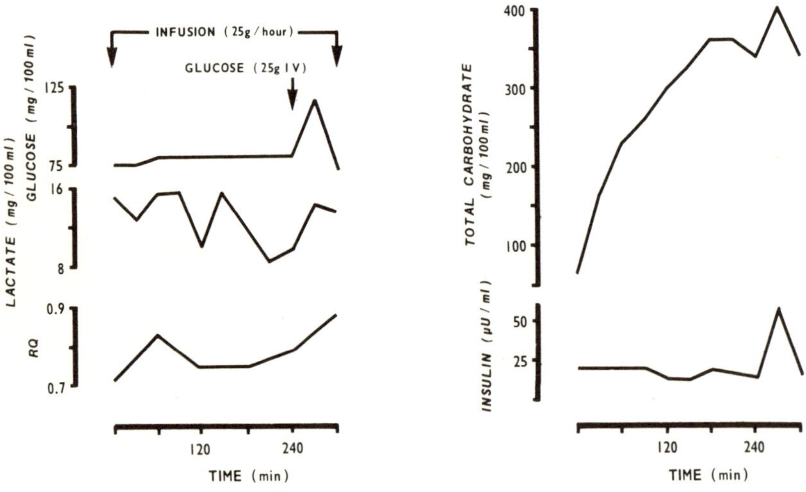

Figure 1 Infusion of Caloreen 25 g/h

Throughout these studies, urinary loss of carbohydrate was marked, usually in excess of 40% of the amount given. Together with the failure to raise the fasting RQ value, this suggested that only a small proportion of Caloreen was rapidly available for metabolic purposes. We conjectured that this may be a function of saccharide chain length and the presence of 1:6 molecular linkages. If this were true then shorter-chain lengths could possibly lead to a readily metabolised material, provided that it was entirely made up of glucose 1:4 links.

The simplest compound in this sequence is, of course, maltose—two glucose residues linked in the 1:4 position. There is some evidence already that maltose is metabolised in man when given intravenously. Studies in the rat[2] showed that if 10 g of maltose was given intravenously to the animal, only 1% was excreted in the urine, whereas 87% of lactose and 63% of sucrose were excreted. Using ^{14}C-labelled sugars only 6.2% of lactose and 7.6% of sucrose were metabolised to expired CO_2; in the case of maltose this was 54.6%. Some 62% of lactose and 68% of sucrose were lost in the urine, but only 4.8% of the [^{14}C]maltose was found in urine. Maltase was found in intestinal mucosa, liver, kidney, brain and in plasma.

Later studies in man showed that after intravenously maltose there was a slight rise in blood glucose levels, a distinct but small rise in serum insulin levels and definite suppression of plasma FFA. The fall in FFA levels ran parallel to the fall after the same weight of glucose[3]. Isotope studies were repeated on human subjects; 61% of the administered ^{14}C was obtained as labelled CO_2, and only 8% was excreted in the urine.

Other workers have not noted an increase in insulin levels following maltose[4]. Maltose, sucrose and lactose have been shown not to stimulate insulin release from the perfused rat pancreas in concentrations up to 400 mg/ml, although glucose produces a brisk response[5].

It has been shown that the peak excretion of [^{14}C]CO_2 is much later after maltose than after glucose, suggesting a rather slower rate of metabolism. Studies in man using ^{14}C labelled material showed that the peak specific activity of expired CO_2 occurred 120 minutes after glucose labelled with ^{14}C, and 235 minutes after maltose. In rats the rate of breakdown to CO_2 could be accelerated by the simultaneous injection of insulin[7].

We therefore repeated our studies as performed with Caloreen, but using maltose as an iso-osmolar 10% solution. After bolus injection we found a small but distinct rise in blood glucose, with a comparable rise in serum insulin. FFA levels fell progressively; lactate levels tended to diminish (Figure 2).

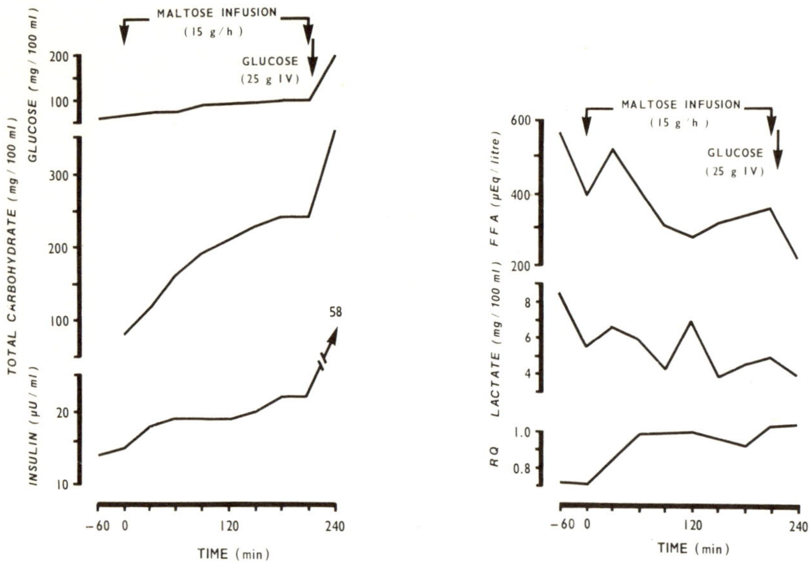

Figure 2 Bolus injection maltose 25 g intravenously

During continuous infusion there were similar changes in insulin and glucose; FFA and lactate levels again fell, and the RQ returned to 1.0. Glucose injection did not produce the sharp changes seen when given during Caloreen infusion (Figure 3).

Further studies using single bolus injections of maltose confirmed the initial findings, but the RQ showed only a modest rise, which may be indicative of the slower metabolism of maltose. FFA and lactate levels fell and glucagon and growth hormone were suppressed.

In all studies, urinary losses were less than 5 g/100 ml. The changes in blood glucose levels were not due to free glucose in the infusions as no glucose was detectable by chromatography of the in-

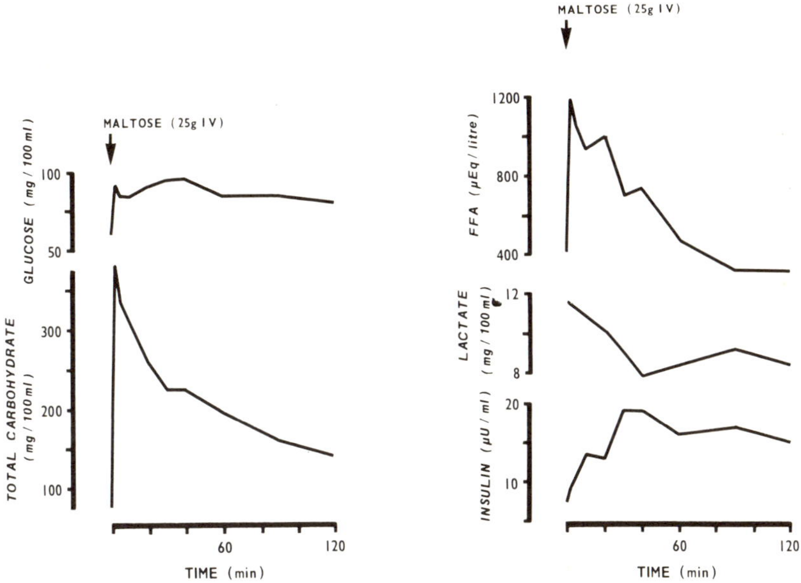

Figure 3 Continuous infusion maltose 15 g/h

fusion solutions. The evidence is clear that maltose given by infusion as an iso-osmolar 10% solution can be metabolised in man as a calorie source, with glucose-sparing effects.

In none of our studies, including the 4-hour infusions of 10% maltose and of 30% Caloreen, did we find thrombophlebitis.

The value of readily metabolised calorie sources at such high iso-osmolar concentrations is obvious. The next step is to seek a 4- or 5-glucose-residue compound with entirely 1:4 links, and to apply similar studies.

REFERENCES

1. Mallick, N. P., Davies, D. and Dobbs, R. J. (1972). *Uremia*, p. 186. (R. Kluthe, G. Berlyne and B. Barton, eds.) (London: Churchill Livingstone)
2. Weser, E. and Sleisinger, M. H. (1967) *J. Clin. Invest.*, **46**, 499
3. Young, J. M. and Weser, E. (1971). *J. Clin. Invest.*, **50**, 986

4. Toyota, T., Kudo, M., Sato, S. and Goto, Y. (1973). *VIIIth Congress of the International Diabetes Federation.* Excerpta Medica International Congress Series, **280,** 93
5. Toyota, T., Ando, Y., Nishimura, H. and Hirata, Y. (1971). *Tohoku J. Exp. Med.,* **104,** 325
6. Young, E. A. and Weser, E. (1974). *J. Clin. Endocrinol. Metab.,* **38,** 181
7. Young, J. M. and Weser, E. (1970). *Endocrinology,* **86,** 426

Discussion

Mr Milner—*Dr Ricketts, you have evidence that in tissue cultures, maltose is metabolised as a source of energy by cells. In what form do you think the maltose is being metabolised?*

Dr Ricketts—*We have done some experiments in which guinea-pig skin, which behaves very like human skin, was grown in tissue culture. A few mg of guinea-pig skin is floated on a drop of guinea-pig serum or buffer solution to provide the nutrient. When we add maltose as nutrient the skin cells take up the maltose and respire and give out carbon dioxide. The volume of carbon dioxide given out can be measured. We looked at the solution to see if any maltose had been hydrolysed to glucose and it had not. I think this shows that the maltose is taken up and metabolised directly by skin cells. If skin cells can do this, surely other cells in the body such as liver cells can also do it.*

Summaries

Metabolic Studies using Infusions of Glucose Polymers and Disaccharides

D. DAVIES and N. P. MALLICK

Intravenous Caloreen given as a bolus to fasting volunteers produced an increase in total serum carbohydrate level and urinary excretion of Caloreen saccharides. The serum insulin level did not rise. Free fatty acid and lactate levels were unchanged. Continuous Caloreen infusion produced similar results and in addition the respiratory quotient remained low. Both insulin and respiratory quotient rose sharply if intravenous glucose was administered. Urine analysis suggested that some of the 1:4 linked saccharides were being metabolised. Maltose (a disaccharide) was given both by bolus and as an infusion; the results suggested that it is utilised for energy. A 1:4 linked glucose polymer containing 4 to 6 glucose molecules might prove to be an effective source of intravenous nutrition.

Etudes Métaboliques lors de la Perfusion lente de Polymères de Glucose et de Disaccharides

D. DAVIES et N. P. MALLICK

L'administration sous forme d'une grosse pilule (bol) de Caloreen intraveineux à des volontaires à jeun provoque un accroissement de la teneur totale en hydrates de carbone du sérum et de l'excrétion urinaire des saccharides du Caloreen. La teneur en insuline du sérum ne s'accroît pas. Les teneurs en acides gras libres et en lactate demeurent inchangées. L'injection continue lente de Caloreen provoque des résultats semblables et de plus le quotient respiratoire demeure bas. La teneur en insuline et le quotient respiratoire s'élèvent brusquement lorsqu'on administre du glucose par voie intraveineuse. L'analyse des urines suggère qu'une partie des saccharides à liaisons 1:4 est métabolisée. On a administré le maltose (un disaccharide) sous forme d'une grosse pilule et d'injection lente; les résultats suggèrent qu'il est utilisé comme source d'énergie. Un polymère de glucose à liaisons 1:4 renfermant 4 à 6 molécules de glucose pourrait constituer une substance efficace d'alimentation par voie intraveineuse.

Stoffwechseluntersuchungen unter Verwendung von Infusionen mit Glukosepolymeren und Disacchariden

D. DAVIES und N. P. MALLICK

Als Bolus an nüchterne Probanden verabreichtes intravenöses 'Caloreen' führte zu einer Zunahme des gesamten Kohlenhydrat-Serumspiegels und der Harnausscheidung der 'Caloreen'-Saccharide. Der Insulin-Serumspiegel stieg nicht an. Die Konzentrationen an freien Fettsäuren und Laktat waren unverändert. Die kontinuierliche 'Caloreen'-Infusion ergab ähnliche Ergebnisse, und zusätzlich blieb der respiratorische Quotient niedrig. Wenn Glukose intravenös verabreicht wurde, stiegen Insulin und RQ steil an. Aus der Harnanalyse folgt, daß einige der 1:4-verknüpften Saccharide abgebaut werden. Maltose [ein Disaccharid] wurde als Bolus und als Infusion verabreicht; aus den Ergebnissen kann geschlossen werden, daß sie zur Energiegewinnung verwertet wird. Ein 1:4-verknüpftes Glukosepolymer, das 4 bis 6 Glukosemoleküle enthält, könnte sich für die intravenöse Ernährung als zweckmäßig erweisen.

Estudios Metabólicos Realizados con Infusiones de Polímeros de la Glucosa y Disácaridos

D. DAVIES y N. P. MALLICK

Caloreen fue administrado en forma de un bolo intravenoso a voluntarios en ayunas y produjo un incremento en el nivel sérico global de carbohidratos y en la excreción urinaria de sacáridos de Caloreen. El nivel de insulina sérica no aumentó. Los niveles de lactato y de ácidos grasos libres permanecieron inalterados. La infusión continua de Caloreen produjo resultados similares, además de ello el cociente respiratorio se mantuvo bajo. Tanto la insulina como el cociente respiratorio aumentaban bruscamente si se administraba glucosa intravenosa. El análisis de orina sugería que algunos de los sacáridos con unión 1:4 estaban siendo metabolizados. Se administró Maltosa (un disacárido) por bolo y por infusión: los resultados sugerían que era utilizada como fuente de energía. Un polímero de la glucosa con unión 1:4, constituido por 4 a 6 moléculas de glucosa puede llegar a convertirse en una fuente efectiva de alimentación intravenosa.

Index